ESSENTIALS OF
MICROBIOLOGY

ESSENTIALS OF
MICROBIOLOGY

K.S. BILGRAMI

D.Phil., D.Sc., FNA, FNASc.
University Department of Botany
BhagalpurUniversity
Bhagalpur-812007(India)

and

R.K. SINHA

M.Sc., Ph.D., FBS, FPSI
Centre for Regional Studies
University Department of Botany
Bhagalpur University
Bhagalpur-812007(India)

CBS Publishers & Distributors Pvt. Ltd.

New Delhi • Bengaluru • Chennai • Kochi • Kolkata • Mumbai • Pune

ISBN: 81-239-0125-9

First Edition: 1992
Reprint: 1997, 2000, 2004, 2005, 2007, 2010, 2015

Published by:
Satish Kumar Jain for CBS Publishers & Distributors Pvt. Ltd.,
4819/XI Prahlad Street, 24 Ansari Road, Daryaganj, New Delhi - 110002
delhi@cbspd.com, cbspubs@airtelmail.in • www.cbspd.com
Ph.: 23289259, 23266861, 23266867 • Fax: 011-23243014

Corporate Office: 204 FIE, Industrial Area, Patparganj, Delhi - 110 092
Ph: 49344934 • Fax: 011-49344935
E-mail: publishing@cbspd.com • publicity@cbspd.com

Branches:
• *Bengaluru:* 2975, 17th Cross, K.R. Road, Bansankari 2nd Stage,
  Bengaluru - 70 • Ph: +91-80-26771678/79 • Fax: +91-80-26771680
  E-mail: cbsbng@gmail.com, bangalore@cbspd.com
• *Chennai:* No. 7, Subbaraya Street, Shenoy Nagar, Chennai - 600030
  Ph: +91-44-26681266, 26680620 • Fax: +91-44-42032115
  E-mail: chennai@cbspd.com
• *Kochi:* Ashana House, 39/1904, A.M. Thomas Road, Valanjambalam,
  Ernakulum, Kochi • Ph: +91-484-4059061-65
  Fax: +91-484-4059065 • E-mail: cochin@cbspd.com
• *Kolkata:* 6-B, Ground Floor, Rameshwar Shaw Road, Kolkata - 700014
  Ph: +91-33-22891126/7/8 • E-mail: kolkata@cbspd.com
• *Mumbai:* 83-C, Dr. E. Moses Road, Worli, Mumbai - 400018
  Ph: +91-9833017933, 022-24902340/41 • E-mail: mumbai@cbspd.com
• *Pune:* Bhuruk Prestige, Sr. No. 52/12/2+1+3/2,
  Narhe, Haveli (Near Katraj-Dehu Road Bypass), Pune - 411041
  Ph: +91-20-64704058/59, 32342277 • E-mail: pune@cbspd.com

Representatives:

• Hyderabad: 0-9885175004	• Nagpur: 0-9021734563
• Patna: 0-9334159340	• Vijayawada: 0-9000660880

Printed at:
J.S. Offset Printers, Delhi

PREFACE

During recent years the study of micro-organisms has provided significant insight about the basic intricacies of the life forms. Many excellent text books and reference works by authors of international repute have come out. Some Indian authors have also produced quite informative account on different aspects of Microbiology. However, there is a lack of concise presentation of knowledge that fulfils the requirement of the curricula of different universities. This book is based on the syllabus prepared and approved by the Curriculum Development Committee of the U.G.C. It has been essentially written to cater to the needs of both undergraduate and post-graduate students of Indian universities.

The book is presented in two major sections : Fundamental Microbilogy (Chapters 1-7) and Applied Microbiology (Chapters 8-11). The first section deals with the forms, structures, functions and classification of micro-organisms (bacteria, mycoplasmas and viruses) in the light of the changing concepts. Chapter 7 concentrates on the genetics of micro-organisms giving an insight into the advanced knowledge of molecular biology. The applied aspects of micro-organisms like industrial microbiology, agricultural microbiology and medical microbiology have been accommodated in section two. A separate chapter on Immunology, Hypersensitivity and Serology has been incorporated to familiarise the students with the intricacies of disease resistance and various laboratory practices. Each chapter is appended with suggested readings to stimulate the students for further study in related disciplines.

We thank all those authors, editors and publishers whose informations provided us base material for the preparation of this text. Our thanks are also due to Mr. Ram Babu for the careful and faithful execution of the drawings and M/S Helpmate (Data Processing Centre), Bhagalpur for computer typing of the manuscript. The CBS Publishers and Distributors and Super Computers of Delhi deserve our special thanks for bringing out this reasonably priced book.

Any suggestion or fruitful criticism by the readers will help us to improve the future editions.

K.S. BILGRAMI

R.K. SINHA

CONTENTS

CHAPTER - 9

MICROBES IN SOIL 113-129

CHAPTER - 10

MICROBES AND DISEASES 130-149

CHAPTER - 11

IMMUNOLOGY, HYPERSENSITIVITY AND SEROLOGY 150-165

x Contents

1

CHANGING FACE OF MICROBIOLOGY

History of microbiology is as fascinating as the history of mankind. Man's concern with microbes was initiated because of the diseases. In the primitive society the most influential and popular man was the witch doctor who was regarded as a combination of prophet, priest, philosopher and physician. Various spiritual and supernatural causes were attributed to diseases - the wrath of God or the possession of demon etc. - but only rarely was a naturalistic hypothesis advanced.

Science of microbiology owes much to Anton Van Leeuwenhoek who first of all introduced microbes (bacteria) to the world. An amateur lens grinder of extraordinary skill and patience, Leeuwenhoek, a resident of small city of Delft, Holland, got interested in observing the tiny creature under his self-made microscope. His objects of study were the teeth scab, leg of flea, head of mosquito, semen, stagnant water, rotten food etc. The **animalcule**, as Leeuwenhoek called them in a series of communications to the Royal Society of London, captured the attention of the contemporary scientists in having detailed and accurate microscopic descriptions of the small creatures. His 39th letter dated September 17, 1683 in which he described the forms in materials scraped from teeth established his genius unequivocally and put him on the highest pedastal of the scientific world. Leeuwenhoek was not a biologist. He may not have been the first man to see bacteria under microscope, but he was certainly the first to describe these forms accurately, in details. He made people realise the importance of microscope as laboratory tool. And his animalcules' which were later named as bacteria received their recognition as a friend as well as foe of the mankind.

BIOGENESIS - ABIOGENESIS CONTROVERSY

Leeuwenhoek's discovery of microbes spurred interest in the origin of living beings. It gave a new insight to the biogenesis- abiogenesis controversy. For long it was believed that living organisms arose *de novo* and life was generated spontaneously from the non-living materials. The contention was ardently advocated by a person not less than Aristotle himself and it took more than a century, before the world was convinced that all living things must have living parents. A group of men in Florence, Italy, formed a society '*Academia del Cemento*' of which Francesco Redi (1626-1698) was an activist. He wanted to disown the theory of spontaneous generation of life. He performed a very simple experiment. He took two jars containing meat, one was covered with a gauze. Attracted by the odour of the meat, the flies laid eggs on the meat and from the eggs maggots developed. The covered jar did not show any sign of life. Redi's experiment was a novice play. But it had tremendous practical significance. It laid the foundation of **Controlled Experiment** in biological science. In order to examine the effect of a particular variable in modern science experiments are performed with one extra control set having all but one variable.

Redi and his followers were vehemently opposed by the contemporary fundamentalists. To give a jolt to the neo-concept, John Needham (1713-1781), an English priest performed an exciting experiment in 1749 in which mutton was heated in a big flask for considerable duration and the mouth of the flask was closed with cork. After a few days small creatures crept into the sealed flask. He opined that had the life arisen from a living parent that would have been destroyed by heating which was not evidenced in his experiment, confirming thereby the concept of spontaneous generation of life . Where was the fallacy ? The fallacy was in the heating of substance. Needham's experiment was repeated by Spallanzani (1764), an Italian priest, with some modifications. Instead of closing the flask with cork, Spallanzani sealed them hermetically in a flame before heating the contents. He boiled the flask for considerable duration and found no sign of life at all. Now it is well known that there are many bacteria which can survive at fairly high temperature and these can be destroyed only when these are exposed for longer duration of heat treatment.

In succeeding years many champions for and challengers of the theory of spontaneous generation figured on the scene, each with a new and sometimes fantastic explanations or experimental evidences. The doctrine received a major blow when Louis Pasteur (1860) took up the challenge to discredit the theory of abiogenesis. Born in the city of Dole in France and starting his brilliant career as a Professor of Chemistry at the University of Lille just at the age of twenty five, Pasteur was entrusted to take up the cause of microbial fermentation by the beverage industries. Encountering opposition from the powerful school of chemists who maintained that fermentation and putrefaction were spontaneous chemical processes and that microbial activity was a effect rather than the cause of observed change, Pasteur ventured to discard the theory. He was convinced that microorganisms floated in the air or dust particles were the main cause of spoilage. He prepared a flask with a long, narrow, gooseneck opening. The nutrient solutions were heated in the flask and the air-untreated and unfiltered, could pass in or out, but the germs settled in the gooseneck and no microbe appeared in the solution. His flask yielded no sign of life, disregarding the theory of spontaneous generation.

The last attempt to revive the concept of abiogenesis was made by Pouchet who published in 1859 an extensive report proving its occurrence. Pasteur - Pouchet controversy whirled the scientific world for sometime but finally Pouchet had to withdraw in favour of the former. Thus came the end of biogenesis-abiogenesis controversy. The final blow to the spontaneous generation doctrine was given by the English physicist Tyndall (1820-1893) in 1877. He conducted experiments in a specially designed box and proved that dust carried the germs. Air containing no dust particle was optically empty i.e. a beam of light passed through it could not be seen whereas each particle in a dust laden atmosphere was clearly visible. He discovered highly resistant bacterial structure, later known as **endospore**, in the infusion of hay. Prolonged boiling or intermittent heating was necessary to kill these spores, to make the infusion completely sterilised, a process known as **Tyndallisation**.

GERM THEORY OF DISEASE

Pasteur's success in solving the problem of fermentation led the French government to ask him to investigate the **pebrin**, a silk worm disease that was ruining the silk industry. After several years of concerted efforts he could isolate the parasite causing the disease and advised the farmers to use only healthy, disease-free worms for breeding stock. Even before Pasteur, strong arguments were placed by different workers for the germ theory of disease. Pasteur also proved by experiments that bacteria were the causes of some diseases. While dealing with the problem of **anthrax**, a disease of cattle, sheep and sometimes human beings, he grew microbes in laboratory after isolating them from the blood of animals that had died of disease. However, the fact that bacteria do produce diseases, was first clearly demonstrated by Robert Koch in 1876. A meticulous German physician, Koch (1843-1910) was also busy with anthrax problem. Pasteur succeeded in cultivating the anthrax bacillus in laboratory but it was Koch who discovered the typical bacilli with squarish ends in the blood of cattle that died of anthrax.

Prior to Koch, Davaine (1863-1868) had also demonstrated rod-shaped objects in the blood and organs of diseased animals. Koch grew these bacteria through a long series of microscopic cultures in his laboratory, examined them microscopically to be sure that he had only one kind of bacterium that could produce typical symptom of the disease in the healthy animals. The bacterium was *Bacillus anthracis*. This was the first evidence that bacterium was the cause of animal disease and a causal relationship existed between the organism and the disease. The result was the establishment of **Koch's postulate** which is as follows :

(i) specific organism must always be associated with a given disease.

(ii) it must be isolated in pure culture.

(iii) when inoculated into a healthy susceptible host it must produce the disease.

(iv) the organism must be reisolated in pure culture from the experimentally infected host.

In order to test the pathogenicity of an organism, Koch's postulates have to be satisfied *in toto*. Using the technique, Koch discovered the bacterium causing tuberculosis in a man and identified it as *Mycobacterium tuberculosis*. He was the teacher who taught his students how to handle the microbe and is popularly known as the **Father of bacteriological technique**.

The imporance of Koch's postulates had tremendous impact on the science of microbiology in the years to come. However, strict adherence to Koch's postulates sometimes led investigators up dead end street. Earlier investigators did not know about viruses and other obligate parasites which are difficult to culture in synthetic media. They were even not familiar with synergism of two or more microorganisms in causing diseases. Today we are much interested in mixed microbial population and its apparent manifestation. The concept of **pure culture** has also changed. It has been substituted with the term **axenic culture**. The former implies genetic purity while the later does not. Moreover, to find a healthy susceptible host is itself ambiguous. With the development of concept of resistance and immunity the susceptibility of the host is defined in different ways. Whatsoever be the limitations of Koch's postulates, it is the tool of circumstantial evidence in the hand of microbiologists to testify the pathogenicity of a given organism.

IMMUNISATION

The treatment or prevention of infectious disease has always been a matter of interest. The best approach for combating infectious disease is **prophylaxis** i.e. disease prevention. Louis Pasteur who had his finger in many scientific pies, became interested in discoveries concerning the cause and prevention of infectious disease. Around 1880 he isolated the germ responsible for chicken cholera. He arranged for a public demonstration that he really had isolated the organism responsible for the disease. He inoculated healthy chickens with his pure cultures but to his dismay, the chickens failed to get sick. Reviewing each step of his experiment, Pasteur found that he had accidently used cultures several weeks old for demonstration, not the fresh one. He repeated his experiments with groups of chickens - one was previously inoculated with old culture and the other was as such. Both groups received bacteria from fresh culture. This time chicken of second group died. This led Pasteur to think that in someway bacteria could lose their ability to produce disease (i.e. loosing their virulence) after growing old and these **attenuated** forms (having decreased virulence) could stimulate the system of body to neutralise the effect of virulent organism. This resulted in developing the concept of **immunisation**.

The importance of immunisation had been realised since the early time. The ancient Chinese deliberately inoculated healthy persons with pustular material from the sores of patients with mild small pox to produce a similar mild disease that would induce long immunity to the disease. The practice continued through the ages and spread to other parts of the world. In the 18th century Lady Mary Wartley Montague attempted to introduce it to England on her return from Near- East. She had to encounter stiff resistance from the people due to old dogmas and conviction. Later on Edward Jenner (1778), a British physician, recognised the immunising procedure for small pox disease. Observing that milkmaids who developed cowpox pustules on their hands and arms after milking cows infected with cowpox, never afterwards had small pox in their lives i.e. they developed immunity for small pox. Jenner deduced that there was relationship between these two diseases. He transferred pustular materials from the infected cow to the skin of human, who shortly displayed the typical sore and later on became immune to small pox when deliberately inoculated with small pox germs. Jenner's experimental significance was realised by Pasteur who next applied this principle to the prevention of anthrax and it worked. He called the attenuated cultures **vaccines** (Vacca=cow) and the process as **vaccination**. We now know that success of Jenner's experiment was due to the similarities in the physical and chemical structure between small pox and cowpox viruses which could produce identical specific blood substance **antibodies** in response to the presence of foreign agents.

Encouraged by the successful prevention of anthrax by vaccination, Pasteur marched ahead towards the service of humanity by making a vaccine for **hydrophobia** or rabies, a disease transmitted to people by bites of dogs, cats and other animals. As with Jenner's vaccination for smallpox, the principle of the preventive treatment of rabies also worked fully which laid the foundation of modern immunisation programme against many dreaded diseases like diphtheria, tetanus, pertussis, polio and measels etc.

The period from 1880 to 1900 was marked by the discovery of various techniques for studying the so called immune reactions both *in vivo* and *in vitro*. Metchnikoff in 1884 proposed the **cellular theory of immunity** involving phagocytic cells (phagein = to eat) present in the blood and tissues of most animals. According to him phagocytes were the body's most important line of defense against infection. The opposing doctrine, the **humoral theory**, put forward by Paul Ehrlich, maintained that soluble substances in the body fluid (e.g. the blood serum) are responsible for immunity. It was observed that cell free serum of certain immune individual was lethal to the bacteria to which they were immune. Serum from such immune individuals reacted with the corresponding bacteria in the test tube and caused them to clump together or **agglutinate**. The controversy between the cellular and humoral theories of immunity raged for several years and was finally resolved when Wright and Douglas showed that although antibody was not necessary for **phagocytosis** it enhanced the rate and extent of phagocytosis.

While all these milestones were being laid in the field of microbiology, an English surgeon, Joseph Lister (1827-1912) was trying to combat the microbes that caused post-operative and wound infection. Lister searched for a way to keep bacteria out of wounds and the incisions made by surgeons. He used a dilute solution of carbolic acid to soak surgical dressings that kept the wounds free from microbial infection and healing took place rapidly. So remarkable was his success that the technique was quickly accepted and the principle of present day **aseptic technique** was established.

CHEMOTHERAPY

Prevention of diseases through immunological responses has its own limitation. Some diseases can strike the person many times because infection fails to induce immunity. However, many diseases can be cured by **chemotherapy** i.e. use of chemicals that selectively inhibit or kill the pathogens without killing the patient. As long back as 1500 B.C. Chinese used mouldy materials for treating boils and for centuries it was known that **quinine**, bark of *Chincona*, was effective against malaria. The modern era of chemotherapy began with Ehrlich's search for **magic bullets** in the first decade of twentieth century. His most outstanding success was the compound 606 which was very much effective against *Treponema pallidum*, causal organism of syphilis in man. The compound was commercially named as **salvarson**. German chemist Gerhard Domagk, about 30 years later, tested the antibacterial activities of various chemicals of sulfonamide and discovered that **prontosil** was active against staphylococcal infection. However, its use was restricted for meningitis, trachoma and kidney infection. These chemotherapeutants were subsequently superseded by the accidental discovery of antibiotics. Antibiotics are substances produced by living orgnisms, which in low concentration, are able to kill or inhibit the growth of other microorganisms. Pasteur and Joubert in 1877 had observed that a common air organism which grew as a contaminant in urine produced a substance that destroyed anthrax bacilli. They, however, did not recognise the practical significance of this observation. In 1889, Vuillemin used the term **antibiose** and `*l*' antibiotic in a French journal for chemical substances involved in bacterial antagonism. The first antibiotic compound that had any commercial possibility was **pyocyanin,** a pigment like substance produced by green pus forming organism *Bacillus pyocyanèus*(syn= *Pseudomonas aeruginosa*). Emmerich and Loew (1890) put this compound in market in Germany which was available till 1930. However, due to its toxicity to tissues, its use was limited to surface application for the control of boils.

The chance discovery of **penicillin** by Alexander Fleming in 1929 was a major break through in the field of chemotherapy which opened a new era of antibiotics. A physician at St. Mary Medical School, London, Fleming was studying the antimicrobial action of varieties of substances including tears and saliva. His laboratory was usually cluttered with old bacterial cultures and he used to keep these cultures for considerable duration for further observation. In some of the petriplates of *Staphylococcus* (the bacterium on which he was working) colonies of mold appeared as laboratory contaminant which developed conspicuous zones of inhibition in the immediate vicinity of the bacterial colonies. The mold was subsequently identified by Charles Thom of USA as *Penicillium notatum* and the metabolite produced by the mold was known as mold juice. Fleming's discovery was little recognised at that time as those were the days of World War II and Fleming had to move to USA to pursue his work. After ten years in 1938, two biochemists of William Dunn School of Pathology, Oxford University, Howard Florey and E.B. Chain joined him and the first human trial of **penicillin** was made in 1941. Subsequently commercial manufacture of penicillin was undertaken in the United States and by 1943 the total production of penicillin was 29 lb costing three million dollar. For their marvellous service to humanity Fleming, Florey and Chain were jointly awarded Nobel Prize. At the same time, Waksman at Rutgers University and others began to search for additional antibiotics from the soil microflora and **streptomycin**, one of the most important antibiotics, was isolated from the broth of *Streptomyces griseus* which was very much effective against *Mycobacterium tuberculosis*.

APPLIED MICROBIOLOGY

Although the greatest interest in microorganisms is related to their role in disease production and their control, they are important to mankind in numerous ways. While Pasteur, Koch and their students were engaged in unravelling the facets of clinical microbiology, some workers in other parts of the world were involved in exploiting these tiny organisms for the welfare of human beings. The basic fact that microorganisms have the potency to convert the raw materials into utilisable products led the foundation of **industrial microbiology**. E.C. Hansen (1842-1909) opened the way to the study of **industrial fermentation**. He developed the pure culture study of yeast and bacteria for **vinegar** manufacture. Similarly Adametz (1889), an Austrian, used pure culture starter for butter production which gave a new dimension to dairy industry. Subsequently microorganisms were aptly exploited for the production of many organic acids of commercial values as well as the pharmaceutical products.

Besides industrial application of microbes, an independent discipline, **soil microbiology**, appeared on the scene in the late 19th century. Soil is the biggest repository of microbes and the dynamism of the soil is largely triggered by the microbial component which ultimately affects the agricultural yield. Biological nitrogen fixation in plants has been one of the most catching subjects for soil microbiologists. It has been known since centuries that legumes enrich the soil. Boussingaults (1838) reported that the favourable action of legumes upon the soil is due to their power to fix atmospheric nitrogen. Frank (1879) demonstrated that nodules in the roots of legumes are formed as a result of inoculation with microorganism. Similarly Hellriegal and Wilfarth (1888) mentioned that legumes took the nitrogen from the air through the agency of bacteria existing in the nodules of their roots. Subsequently Beijerinck, a famous Dutch microbioligist and Winogradsky, the Russian scientist, isolated many organisms from the soil responsible for **nitrogen fixation** and worked out the mechanism of symbiotic and asymbiotic nitrogen fixation.

DISCOVERY OF VIRUS

Diseases like small pox and yellow fever caused by viruses have been recognised since centuries and even plant virus diseases such as potato leaf roll and the ornamental variegation known as 'tulip break' are known for the last several centuries. Pasteur suspected that the cause of hydrophobia was a submicroscopic entity that could not be cultivated outside the animal body. However, the first disease clearly demonstrated to be produced by viruse was tobacco mosaic. Mayer (1886) demonstrated its transmissibility by mechanical inoculation with sap of infected plants. Iwanowsky (1892) reported the transmission of tobacco mosaic by sap filtered through bacteria proof filter. The report, however, went unnoticed. In 1899, Beijerinck succeeded in proving the serial transmission of disease in bacteria free filtrate in which no microscopic organism could be detected. He described this agent of tobacco mosaic as *"contagium vivum fluidum"* (= an infective agent which reproduced but was in a state of dispersion). Similar findings were made with several animal viruses and the existence of descrete particles or elementary bodies as the infective agent was demonstrated microscopically. After the discovery of electron microscope in 1930 structural details of viruses were studied by different workers. Stanley (1935) successfully crystallised rod shaped virion of tobacco mosaic virus. For his valuable contribution to the knowledge of the nature of viruses and crystallisation of virus protein. Stanley and Northrup were honoured with Nobel Prize in 1946. Not only plants and animals are subjected to virus infections, many bacteria are also parasitised by viruses. A British investigator Twort discovered for the first time such bacterial virus in 1915. Subsequently D'Herelle in 1917 observed that the feces of a bacillary dysentry patient contained an agent which on addition to the broth culture of the dysentry organism killed and apparently dissolved the bacteria within few hours. D'Herelle referred to the agent as **bacteriophage**.

MICROBIOLOGY IN TWENTIETH CENTURY

The above few illustrations clearly reveal that by the end of 1900, science of microbiology grew up to the adolescence stage and had come to its own as a branch of the more inclusive field of biology. In the years to follow its horizon widened considerably and the micro-organisms were picked up as ideal tools to study various life processes and thus an independent discipline of microbiology, **molecular biology** was born. The relative simplicity of the microorganism, their short life span and the genetic homogeneity provided an authentic simulated model to understand the physiological, biochemical and genetical intricacies of the living organisms. The list of those who contributed to the development of microbiology is far too long to recite here in its entirety. Here we shall highlight briefly only few of the important microbiological accomplishments (see table 1.1 also).

The Nobel Prize winning research on culturing of poliomyelitis virus by Enders, Robbins and Weller provided the basis for the development of the vaccine used for the prevention of poliomyelitis by Salk and his colleagues in 1954. The physiology and metabolism of microorganisms were unravelled by Fritz Lipman and Hans Krebe who in 1953 were awarded the Nobel Prize. In 1958 Joshua Lederberg, George Beadle and Edward Tatum were awarded Nobel Prize for their discoveries that genes act by regulating specific chemical processes at the cellular level and that genes in bacteria are organised in specific ways which can be altered by process called **recombination**. The well known concept of **'one gene one enzyme'** deduced by Beadle and Tatum provided clues needed for understanding the process of heredity in higher organisms. Ochoa and Kornberg isolated and synthesised the enzyme responsible for production of ribonucleic acids, RNA and DNA, that carry hereditary information for which they received Nobel Prize in 1959. The Nobel Prize in physiology and medicine for 1968 was shared by Robert W. Holley, Har Govind Khorana and Marshall. W. Nirenberg for their contribution to the understanding of the genetic code and its function in protein synthesis. Holley at Salk Institute determined the chemical structure of tRNA. Khorana, at the University of Wisconsin proved by synthesis of nucleic acids having a known sequence of bases that the genetic code is made up of triplets. Nirenberg, at the National Institute of Health, deciphered the genetic code that prescribes the manufacture of proteins. The 1969 Nobel Prize in medicine and physiology was awarded to Max Dalbruck, Alfred D. Hershey and S.E. Luria for studying the replication mechanism and genetic structure of bacteriophage. In 1972 two immunologists Gerald M. Edelman of the U.S.A. and Rodney R. Porter of U.K. shared the Nobel Prize for contribution to the knowledge of the chemical structure of antibodies. Albert Claude, George E. Palade and Christian R. Duve jointly received this prestigious prize of 1974 for introducing differential centrifugation for the isolation of cell parts in order to study the structure and chemistry of individual cell which led to the discoveries of ribosome and lysosome. The following year in 1975 R. Dulbeco, H.M. Temin and David Baltimore of the U.S.A. were awarded the Nobel Prize for researching the interaction between tumor virus and genetic material of the cell. In 1976 Gajdusek and Blumberg did research leading to Nobel Prize for a test to show hepatitis virus in donated blood and to an experimental vaccine against the disease. Two years later in 1978 Arber, Smith and Nathans were jointly awarded this prize for discovery of **restriction enzyme** and their application to the problems of molecular genetics. Nobel Prize for chemistry in 1980 went to Paul Berg, Water Gilbert and Frederick Sanger for development of a rapid way to determine the chemical make up of DNA. A distinct relationship between EB virus and Burkitt's lymphoma was established by Epstein and Barr showing direct involvement of virus in cancer for which they were awarded the Nobel Prize of 1982.

From the foregoing it is evident that the history of microbiology is an unfinished epistolary record. The secret of nature always allures the genius and it is the inquistiveness of man that prompts him to lay the milestones to unfold the facets of life with the help of these tiny organisms. The future awaits diligent investigations from the prepared minds.

Table 1.1

IMPORTANT LANDMARKS IN MICROBIOLOGY - AN OVERVIEW

Year	Event
1676	Anton van Leeuwenhoek discovered 'little animalcule'.
1688	Francesco Redi published book on spontaneous generation of maggots from putrid flesh.
1720	Bradley's germ theory.
1765	Spallanzini showed that 'organic molecules' are distinct organisms.
1770	Hill introduced new methods of staining and preserving specimen for microscopic study.
1796	Edward Jenner inoculated James Phipps with cowpox.
1802	First use of word 'biology' (Treveranus).
1835	Bassi's theory of 'living contagion' in silk worm disease.

IMPORTANT LANDMARKS IN MICROBIOLOGY - AN OVERVIEW (contd.)

1838	Liebig established biochemistry.
1844	Bassi asserted smallpox, bubonic plague, syphilis, spotted fever due to living parasites.
1850	Davaine asserted that 'anthrax' was due to 'bacteroides' which he saw in the blood of dead sheep.
1857	Pasteur demonstrated that lactic acid fermentation is due to living organism.
1864	Pasteur invented pasterurization for wine.
1867	Joseph Lister published work on antiseptic surgery.
1876	Robert Koch gave three day demonstration of his work on anthrax.
1877	Koch described technique of fixing, staining and photographing bacteria. Also discovered *Bacillus anthracis* as cause of anthrax.
1879	Albert Neisser discovered *Neisseria gonorrhoeae* as cause of gonorrhea.
1881	Koch worked out method of culturing bacteria on gelatin.
1882	Koch discovered tubercle bacillus and enunciated 'Koch postulates'. Metchnikoff proposed phagocytic theory-cellular theory of immunity.
1885	Pasteur inoculated Joseph Meister for Rabies. Theodor Escherich discovered *E. coli.*
1898	Beijerinck discovered and named tobacco mosaic virus.
1902	Landsteiner investigated agglutination when blood from different human donors was mixed.
1903	Wright and other discovered opsonin (i.e. antibodies) in blood of immunised animals.
1906	Schaudinn and Hoffman discovered *Treponema pallidum* as cause of syphilis. Von Wassermann developed test for syphilis
1912	Ehrlich demonstrated first chemotherapeutic agent for bacterial disease.
1915	D'Herelle and Twort independently showed the existence of bacteriophage.
1923	Landsteiner observed M and N factor in blood.
1926	Ultracentrifuge (Svedberg).
1929	Alexander Fleming discovered Penicillin.
1932	Sir Hans Krebe described the Citric Acid Cycle.
1933	First electron microscope (Ruska).
1935	Stanley crystallized virus.
1941	Beadle and Tatum established 'one gene-one enzyme' theory.
1952	Waksman discovered Streptomycin, the antibiotic effective against tuberculosis.
1953	Lipman discovered Coenzyme A and its importance for intermediary metabolism. Lederberg and Zinder discovered transduction. Crick and Watson proposed a structure for Deoxyribonucleic acid - a double spiral.
1954	Ender, Weller and Robbins discovered poliomyelitis viruses in culture of various types of tissues.
1957	Virus structure determined. Inferferon discovered by Issac and Lindeman.
1958	Lederberg made discoveries concerning genetic recombination and the organisation of the genetic material of bacteria.
1959	Kornberg and Ochoa discovered the enzymes responsible for producing artificial DNA and RNA.

1961	Nirenberg using artificial DNA, synthesised a protein molecule.
1962	Crick, Watson and Wilkins made discoveries of molecular structure of nucleic acid and its significance.
1966	Rous discovered tumor inducing virus.
1968	Holley, Khorana and Nirenberg investigated the genetic code and its function in protein synthesis.
1969	Delbruck, Hershey and Luria discovered replication mechanism and genetic structure of viruses.
1972	Edelman and Porter awarded Nobel Prize for discovery of chemical structure of antibodies.
1974	Claude, Duve and Palade described structural and functional organisation of the cell.
1975	Baltimore, Dulbeco and Temin researched the interaction between tumor virus and the genetic material of the cell.
1976	Gajdusek and Blumberg devised a test to detect hepatitis virus.
1978	Arber, Smith and Nathans discovered restriction enzymes and their application.
1979	Henle identified first virus regularly associated with human cancer.
1980	Paulberg, Gilbert and Sanger awarded Nobel Prize for developing rapid way to determine chemical make up of DNA.
1982	Epstein and Barr showed relationship between EB virus and Burkitt's lymphoma.
1983	Jean Dausset of France and Barbara McClintock of USA discovered the transposable genetic element.
1984	Cesar Milstein developed monoclonal antibody technique.
1987	Kohler, Niels and Susumu-Tomegawa were jointly awarded Nobel Prize for researches in immunology and generation of antibody diversity.

SUGGESTED READINGS

Bulloch, W. (1938) The History of Bacteriology, Oxford University Press, London.

Carpenter, P.L. (1967) Microbiology W.B. Saunders Company Philadelphia and London.

De Kruif, P. (1966) Microbe Hunters, Harcourt Brace Jovanovich, New York.

Dobell, C. (1960) Anton van Leeuwenhoek and his "Little Animals". Dover Publications, New York.

Gabriel, M.L. and S. Fogel (ed.) (1955) Great Experiments in Biology, Prentice - Hall, Engle Wood Cliffs. N.J.

Lechevalier, H. and M. Solotorovsky (1974) The Centuries of Microbiology, Dover Publication, New York.

McKane, L. and J. Kandel (1986) Microbiology: Essentials and Applications, McGraw Hill International Edition, Life Science Series, New York.

Pelczar, M.J., R.D. Reid and E.C.S. Chan (1972) Microbiology, Tata McGraw Hill Publishing Company Ltd. New Delhi.

Porter, J.R. (1976) Anton von Leeuwenhoek. Tercentanary of His Discovery of Bacteria, Bacteriological Review 40(2): 260-269.

Postgate, J. (1975) Microbes and Man, Penguin Books, Baltimore.

Taylor, G.R. (1963) The Science of Life, McGraw Hill, New York.

2

THE MICROBIAL WORLD

The world of micro-organisms (*micro* = small; organism = living thing) is beautiful and fascinating. It comprises those small living organisms which are microscopic in nature. Generally these are unicellular organisms that include protozoa, algae, fungi, bacteria, cyanobacteria (blue-green algae) and viruses*. Some molds and several algae are multicellular also. Viruses are submicroscopic acellular entities, yet they are considered as microbes because they have many characteristic features of organisms. In the succeeding chapters we have also used the word microorganisms for all the organisms of the above six categories.

Until the last century, living organisms were classified in two broad groups i.e. **animals** and **plants** due to their obvious differences in form, constitution, and function. The differences were mainly in their mode of nutrition - the former being **autotrophic** and the latter **heterotrophic**. After the introduction of bacteria and other unicellular organisms a third division of living organisms, **protista** (primary or archaic life forms) was established by Haeckel (1866). The kingdom protista contains those organisms that are differentiated from plants and animals by their lack of morphological specialisation. The kingdom has been further subdivided into two clearly differentiated groups on the basis of their cellular structure. The higher protists resemble plants and animals in their cell composition; they are **eukaryotic** that include algae, fungi and protozoa. The lower protists include the bacteria and cyanobacteria; they are **prokaryotic** and their cellular structures are quite different from those of other organisms. A four kingdom system classification is presented in Table-2.1 that exhibits demarcation between prokaryotes and eukaryotes as well as micro- and macro-organism.

PROKARYOTES AND EUKARYOTES

For long it was believed that the structural arrangement within bacteria were the same as those in higher plants and animals. However, with the advancement of electron microscopy it was found that there was a striking difference between the bacterial cell and that of higher organisms. Such differences formed the basis for classifying cells into two categories - eukaryotes (*eu*. true; *karyote*, nut, nucleus) and prokaryotes (*pro*, before). Eukaryotic cells contain true nucleus surrounded by a membrane which separates it from the cytoplasmic contents whereas the prokaryotic cells do not possess any membrane around the nuclear material. Instead their hereditary information is suspended in a portion of cytoplasm called the **nucleoid** which is well distinguished in the nuclear region. The prokaryotes are regarded as relics from the earliest time of biological evolution and the development of eukaryotes represent the greatest discontinuity in the evolution of living organism. The distinction between eukaryotes and prokaryotes are summarised in Table 2.2.

Table 2.1

FOUR-KINGDOM CLASSIFICATION OF LIVING ORGANISMS

PROKARYOTIC CELL	I. Kingdom : **Prokaryotae** 　(a) Bacteria 　(b) Blue green algae	MICROORGANISM
	II. Kingdom : **Protista** 　(a) Protozoa 　(b) Algae 　(c) Fungi	
EUKARYOTIC CELL	III. Kingdom : **Plantae** 　(a) Mosses 　(b) Ferns 　(c) Cone-bearing plants 　(d) Flower bearing plants	MACROORGANISM
	IV. Kingdom : **Animalae** 　(a) Sponges 　(b) Clams 　(c) Insects 　(d) Worms 　(e) Birds 　(f) Mammals	

In prokaryotes DNA exists as a closed circular molecule in the cytoplasm. The single bacterial chromosome contains all the information necessary for reproduction of the cell. In addition to this there may be one or more small circular DNA molecules, called **plasmid**; which are, however, dispensible. Prokaryotic ribosomes are relatively small. There is little morphological differentiation among prokaryotes. Such uniformity in shape, however, is accompanied by a remarkable diversity and flexibility in metabolic properties. Such physiological versatility and flexibility, high rate of growth and synthesis, the simple architecture of the cell and uncomplicated structure of the genetic material have made the prokaryotes the preferred experimental objects of general biology over the last three decade or more.

EUKARYOTIC VERSUS PROKARYOTIC CELL STRUCTURE

Until the perfection in electron microscopy our knowledge of detailed cell structure was based mainly on direct visual observation and light microscopy. Extension of light microscopy to the ultraviolet range and eventually electron microscopy have brought dramatic change in our concept of cell structure. Coupled with these optical observations, the biochemical preparative methodologies have also contributed in the elucidation of function and structure of cell, their organelles and components. These investigations have clearly demarcated the eukaryotic and prokaryotic cells in distinct groups.

EUKARYOTIC CELL

The Nucleus : The structure of the nucleus and the type of nuclear division are the most striking and fundamental differences between eukaryotic and prokaryotic cells. The nucleus is separated from the rest of the cell by nuclear membrane and contains chromosome, structure responsible for transmitting properties to offsprings. All eukaryotic cells contain more than one chromosome (in contrast to prokaryotes, which have only one chromosome). The genetic material of chromosome is deoxyribonucleic acid (DNA). The nucleus contains a small body called the **nucleolus** which manufactures structural components used in protein synthesis.

Table 2.2

DIFFERENCE IN PROKARYOTIC AND EUKARYOTIC CELL.

FEATURE	PROKARYOTES (Nucleoid)	EUKARYOTES (True nucleus)
Group	Bacteria, blue- green algae	Algae, fungi, protozoa, plants and animals
Size range of organism	1-2 x 1-4 μm or less	Greater than 5 μm
Nuclear Region		
Location	Nucleiod, Chromatin body or nuclear material	Nucleus, Mitochondria, Chloroplast
Structure of nucleus	Not bounded by nuclear membrane	Bounded by nuclear membrane
Number of chromosome	One, circular chromosome	One or more, linear chromosome
Chromosome	Chromosome does not contain histones No mitotic division Nucleus absent, Functionally related genes may be clustered	Chromosomes have histones Mitotic division Nucleus present Functionally related gene not clustered.
Cytoplasmic nature and structure		
Cytoplasmic Streaming	Absent	Present
Mesosome	Present	Absent
Ribosome	70 S in cytoplasm	80 S, arranged in endoplasmic reticulum, 70 S in mitochondria and chloroplast
Mitochondria	Absent	Present
Chloroplast	Absent	Present
Golgi apparatus	Absent	Present
Endoplasmic reticulum	Absent	Present
Surface Layer		
Cytoplasmic membrane	Generally donot contain sterol	Sterol present
Cell wall	Peptidoglycan as component	Peptidoglycan absent
Flagella (if present)	Simple fibril, composed of flagellin	Multifibrilled with 9 +2 microtubules
DNA base ratio as mole % of guanine +cytosine (G + C%)	28 to 73	about 40

In eukaryotic cells nuclear division occurs by mitosis which serves two functions. (i) the accurate replication of the genetic material which eventually become visible in the lengthwise separation of the duplicated chromosome and (ii) the equal distribution of two complete sets of chromosome to each of the two daughter nuclei. The exact mechanism of duplication of chromosome is not fully understood, though much of the process of DNA replication is elucidated in prokaryotes.

All the higher plants and animals undergo a nuclear rearrangement as part of sexual replication. At fertilisation the gamets and their nuclei fuse to give rise to the zygote. The male and female nucleus contributes equal number (n) of chromosome during fertilisation. The zygote nucleus, therefore, has two sets (2n) of chromosome (and genome). Thus whilst the gametes are haploid, all somatic cells are diploid. For the next sexual generation, therfore, the normal diploid (2n) chromosome complement has to be reduced to haploid number (n).

The eukaryotic chromosome consists of DNA strands associated with numerous proteins, some of these are basic proteins known as **histones**. Histones and DNA are apparently associated in a high orderly manner, forming **nuclosomes**, which are regarded as subunits of chromosomes. Besides nucleus, mitochondria and chloroplasts of the eukaryotic cells also contain DNA that carry genetic information for the cell.

The Cytoplasm : The eukaryotic cell is filled with a jelly like fluid called the cytoplasm. This cytoplasm remains in constant motion, a phenomenon termed **cytoplasmic streaming**. Proteins and dissolved nutrients are transported through the cytoplasm as a result of this movement. Cytoplasm contains organelles which are intracellular structures bound by membranes. Membrane bound organelles include the mitochondria, the chloroplasts and the vacuole. In addition, an elaborate membrane network, the **endoplasmic reticulum**, is also present in the cytoplasm.

Mitochondria : Most of the energy for cell function is produced in organelles called mitochondria (singular, mitochondrion). Energy is generated in mitochondria by respiration. They are pleomorphic lipid rich structures which consist of two membrane, an external one and a much folded inner one constituting **cristae**. The infoldings contain the component of the electron transport chain and ATP'synthetase.

Chloroplast : The cells of algae and green plants contain chloroplasts in addition to mitochondria. Chloroplast membranes contain chlorophyll, a green pigment used for photosynthesis. The inner membrane of chloroplast (**thylakoids**) are the sites of photosynthetic pigments and components of photosynthetic electron transport. Like mitochondria chloroplasts have their own DNA and ribosome and are also self replicating within the cytoplasm.

Endoplasmic Reticulum : The endoplasmic reticulum is an internal membrane network that extends throughout the cytoplasm from the outer cell membrane to the nucleus. It also forms the nuclear membrane which surrounds the nucleus and contains numerous pores to allow easy transfer of nucleic acid, proteins and metabolites between the nucleus and cytoplasm. Part of the endoplasmic reticulum has many ribosomes attached to it and this is known as the rough or granular endoplasmic reticulum.

Ribosome : Ribosomes are the sites of protein synthesis. Besides being present on membrane structures, these are largely found in cytoplasm. The ribosomes attached to the endoplasmic reticulum and those that are free in the cytoplasm of the cell both belong to the 80 S type.

The Golgi Complex : A special membranous organelle of the animal cell is the Golgi apparatus. Similar organelles found in the plant cells are called **dictyosomes**. They consist of bundle of flattened membrane vesicle, called **cisternae**. Lysosomes are manufactured by stack of flattened membrane disks. Both the golgi apparatus and dictyosomes have important function in secretion of many enzymes that are synthesised on the cisternae and collected inside them.

Vacuole : Many eukaryotic cells form intracellular sac called vacuoles. For example, many protozoa feed by **phagocytosis** engulfing particles into food vacuoles. Other sacs in the cytoplasm, called **lysosome**, are filled with enzymes which help digest engulfed particles.

A characteristic property of eukaryotic cell is the ability to take up solid as well as liquid nutrients from the environment. The uptake of solid particle is known as **phagocytosis** (e.g., leukocytes in the blood). Fluid uptake is called **pinocytosis** (e.g., entry of bacteriophage inside the host). Both kinds of uptake of extracellular material are known collectively as **endocytosis**. The capacity of eukaryote to take up particles is of fundamental biological importance. In the normal course of events, a solid particle, taken up by an amoeba via phagocytosis would be completely digested and assimilated. However, in some cases such endocytosis results in intracellular symbiosis; the best known example of this is leguminous root nodules with rhizobia (Chapter IX). The capacity of eukaryotic cells for endocytosis supports the hypothesis for the endosymbiotic origin of mitochondria and chloroplasts.

Surface layer : Cytoplasmic content in all eukaryotic cell is surrounded by cell membrane or plasma-membrane. It determines the selective transport of the molecules between the external medium and the interior of the cell. Because of this selectivity cells may accumulate nutrients and dispose of toxic waste products. Both eukaryotes and prokaryotes contain same basic structure of membrane composed of double layer of phospholipids with protein dispersed throughout the structure. The eukaryotic cell membrane typically contain sterol (cholesterol in human and animal, and ergosterol in fungi). In fungi and algae the cell membrane is surrounded by a cell wall which confers rigidity and shape to the cell. The cell walls in most eukaryotes are composed of polysaccharide. In plants and most algae the polysaccharide is cellulose whereas in fungi it is **chitin.** In a few species, the cell wall may by surrounded by an additional layer, the **capsule**, which under circumstances, protects the organism from destruction. A typical example is the fungus (yeast) *Cryptococcus neoformans.*

Organelles of locomotion : Eukaryotic cells move from one place to another with the help of accessory organelles called **flagella** or cilia. An amoeba (a protozoa) moves by sending out extension of their surface, the **pseudopods.** Flagella (singular, flagellum) are long filament that whip back and forth, propelling the cell forward. Cilia are shorter than flagella but are almost identical in morphology and chemical composition. Transverse sections show nine peripheral double filaments with two central single filament (the 9 + 2 pattern). They are surrounded by the cytoplasmic membrane. The flagella are inserted in the outer layer of the cytoplasm, via a basal plate or 'blepharoplast' which itself arises from a self duplicating organelles, the **centriole.**

EUKARYOTIC MICROBES

Eukaryotic microorganisms comprise three groups - algae, fungi and protozoa - that are distinguisable from each other due to their characteristic features. A general account of these microbes is given below. For detailed discussion, the students are advised to consult the selected readings given in the end of this chapter.

ALGAE: (singular, alga) are a diverse group of organisms ranging from microscopic single cell to large multicellular seaweeds. They are the simplest chlorophyll containing plants.

Algae are found almost every where on earth, from the tropics to arctic region. They are principally aquatic. Many species live in damp soil, some grow on rocks and some in hot springs at temperature as high as 90° C. Aquatic forms of algae constitute the phytoplankton of ecosystem.

Algae have a wide range of shapes and sizes. They vary in shape from spheres to rods, clubs, spirals and irregular forms. Multicellular species exhibit great variations in form and complexity including membranous colonies; filaments grouped singly or in clusters with individual strands that may be branched or unbranched. Algal cell may be eukaryotic or prokaryotic. In most species the cell wall is thin and rigid. Cell walls of **diatoms** are impregnated with silica; walls of blue green algae (prokaryote) contain peptidoglycan and diaminopimelic acid, as bacteria do. The motile algae such as *Euglena* have flexible cell membranes called periplasts. The cell walls of many algae are surrounded by a flexible gelatinous matrix secreted through the cell wall. Cell contains a discrete nucleus except in blue-green. Other inclusions are starch grains, oil droplet and vacuole. With an exception, *Prototheca*, all algae are photosynthetic and are easily recognised by the presence of chloroplast within the cytoplasm. The chloroplasts may be ribbon-like, bar-like, net-like, or in the form of discrete disks. The chloroplast contains **chlorophyll** and other pigments. There are three kinds of photosynthetic pigments in algae; chlorophylls, carotenoids and phycobilins. There are five chlorophylls : *a, b, c, d* and *e*. Chlorphyll *a* is present in all algae, carotenoids are of two types: **carotenes** and **xanthophylls**. Phycobilins are water soluble pigments whereas chlorophylls and carotenoids are lipid soluble.

The mobility of algae is achieved by means of flagella occurring singly, in chain or in cluster at the anterior or posterior ends of the cell. Flagella may be whiplash (cylindrical and smooth), tinsel (cylindrical with hair like appendages) or strap-like. Some algae are non-motile and are carried by waves and currents of water. These are normally referred to as **phytoplanknton**.

Algae reproduce either sexually or asexually. Asexual reproduction includes single cell division or fission. However, most of the asexual reproductions in algae are accomplished by unicellular spores - the motile spores are known as **zoospores** and the non-motile ones are the **aplanospores**. All the three forms of sexual reproduction viz; isogamous, anisogamous and oogamous are met in algae which are accomplished by the fusion of sex cells (**gametes**) blending the nuclear materials of the opposite sex. Accordingly the plants may be unisexual (**dioecious**) or bisexual (**monoecious**).

Except blue-green algae, the above characteristics are common in almost all other algae and hence the blue-green algae are considered to be quite different from other members by many microbiologists and in **Bergey's Manual of Determinative Bacteriology**, 8th edition (1974) these are classified as cyanobacteria in the kingdom Prokaryotae.

Economic importance of algae is manifold. Some blue-green algae enrich soil fertility. In many countries these are used as biofertilizers. The yellow pigment, carotene, found in many algae is a precursor of vitamin A. Other algae synthesize vitamin D. Green algae contain good amount of vitamins B, C and K. Many species of algae are used as food. *Chlorella* is widely used as food for human and domestic animals. These are good source of **single cell protein** (see Chapter 8). In aquatic system many algal species react sharply to the environment and can be used for biomonitoring of pollution. Some antibiotics are also extracted from algae. Few algae are pathogenic to man and other mammals. For example *Prototheca* causes **protothecosis**. Several species are parasite on higher plants, e.g., the green alga *Cephaleuros* attacks leaves of tea, coffee, pepper, mango etc. Some planktonic algae produce toxins which are lethal to fish and other animals. *Gymnodinium, Gonyaulax* produce neurotoxin; *Microcystis* and *Anabaena* are known to be toxic to birds and mammals.

FUNGI : The name fungus is derived from its most obvious representative, the mushroom (Greek, *mykes*; Latin, fungus). They are eukaryotes and like plants possess a cell wall, liquid-filled intracellular vacuoles, microscopically visible streaming of the cytoplasm and (almost universal) lack of motility. However, they do not contain photosynthetic pigments and are **chemoorganoheterotrophs.**

The vegetative body is a thallus. It is composed of microscopic tubular filaments called hyphae (singular, hypha). The filaments or hyphae consist of a cell wall and cytoplasm with its inclusions. The hyphae may be without cross walls (in the lower fungi) or divided into cells by septa (in the higher fungi). However, even in the septate hyphae the cytoplasm of the cells is continuous maintaining continuity through a central pore in the septum. The total of the hyphal mass of fungal thallus is called as **mycelium** (plural, mycelia). Most fungi are filamentous; some are unicellular e.g., the common yeast. The mycelium may form a loose meshwork or a compact tissue as in the mushroom.

In most fungi every part of the mycelium has the potential for growth (elongation). Reproduction is accomplished by two common methods: sexual and asexual. Most fungi can reproduce in both ways. Asexual reproduction is mostly by **budding, fragmentation** or **spore formation**. The last one is the most widely distributed and most highly differentiated method. The spore producing structure is called as **sporophore**. The terminal portion of the sporophore forms a sac called **sporangium** and is termed as sporangiophore. In the lower fungi, sporangia are often motile by means of flagella and are called zoospores. The non-motile **sporangiospores** are referred to as **aplanospores**. When the spores are borne free, the sporophore is termed as **conidiophore** and the spores as conidia (singular, conidium). Since conidia vary widely in shape, size, colour etc. they serve as good taxonomic criteria for fungi. In some genera more than one type of conidia (microconidia and macroconidia) are formed from the same thallus.

The asexual reproduction characteristic of yeast (budding fungi) is budding; the mother cell forms an outgrowth which receives daughter nucleus, where upon the nucleated outgrowth is purched off as a **bud**. It is the simplest and most primitive type of asexual spore, found in the true yeast *Saccharomyces cerevisiae.* **Blastospores** are thin walled sprout cells formed by budding (e.g; *Candida*); **chlamydospore** is round, thick-walled, formed terminally, intercalary or laterally on the hyphae or conidia (e.g., *Fusarium*); **arthrospore** or **oidium** is formed by segmentation of the hyphae into somewhat rectangular spores (e.g., *Endomyces*).

Like other eukaryotes, sexual reproduction in fungi is attained by union or conjugation of two nuclei of different strains. It is usually divided into three phases. The initial phase is **plasmogamy** i.e. fusion of two protoplasts. The resulting cell has two nuclei. This nuclear pair or **dikaryon** may not fuse immediately but may persist in the dikaryotic stage for varying length of time. The second stage is the fusion of the two haploid nuclei, i.e. **karyogamy** resulting in diploid nucleus of the zygote. Following karyogamy, meiosis occurs restoring the haploid state of the nucleus. These three stages, plasmogamy, karyogamy and meiosis may occur in immediate sequence in some fungi or may be phased development.

In the lower fungi, sexual reproduction is initiated by the formation of gametes (sex cells). When the gametes of the male and female parents are morphologically indistinguishable, they are called **isogametes**. The gametes are formed inside the gametangia; the male gametangia are called the **antheridia** and the female ones the **oogonia**. In the lower fungi, especially in aquatic forms, both gametes are usually motile (**planogametes**) and fuse outside the gametangia. In the oomycetes, only the male gamete is motile. Zygomycetes are characterised by gametangiogamy, the fusion of whole, multinucleate gametangia into a **coenozygote**. When the male and female gametangia originate from the same vegetative body, the organism is referred to as **homothallic**. In **heterothallic** fungi the thalli are either male or female. Homothallic fungi are usually self fertilising (**autogamous**). However, in some homothallic forms, no self fertilisation occurs due to some physiological inhibitory mechanism which is due to **incompatibility**. In Ascomycetes the fusion of two nuclei takes place in penultimate cell of a specialized hypha known as **ascogenous hypha** which after fusion forms a sac like structure known as **ascus** (plural, asci). The haploid spores produced in it are called as ascospores. Asci are produced inside as specialized hyphal bodies called as **ascocarp** which are classified as **cleistothecium, perithecium** or **apothecium** on basis of their morphology. Sexual spores in the Basidiomycetes are known as basidiospores which are usually four in number developed on the club shaped structure called **basidium**. In many species the basidia and their basidiospores are organised in highly developed structures known as mushrooms, puffball and bracket-fungi.

Fungi constitute the decomposer segment of an ecosystem that live as parasite or saprophyte. As parasite they cause many important diseases of plants, human beings and animals. Some molds are known to produce mycotoxins that are highly toxic and in some cases carcinogenic as well. In contrast, many fungi are important in industrial fermentation for the manufacture of wine, production of antibiotics e.g; penicillin), vitamins and organic acids. These have also become important tools for the physiologists, biophysicists, geneticist and biochemists for the study of many fundamental life processes.

PROTOZOA : The name protozoa was derived from the Greek word (*protos* = first) and (*zoon* = animal). These are eukaryotic organisms that occur as single cell or in colonies. In the colonial forms, the **individual cell (organisms) is either joined by cytoplasmic threads or embedded in a common matrix.**

The shape and size of protozoa vary considerably. *Leishmania donovani*, the cause of human disease **kala-azar**, measures only 1-4 µm whereas *Amoeba proteus* measures about 600 µm; some species measure upto 7 c.m. Like other eukaryotes, the protozoan cell also consists of cytoplasm, cell envelope and nucleus. The cytoplasm is more or less homogenous comprising globular protein molecules loosely linked together to form a three dimensional molecular framework. Pigments are diffused throughout the cytoplasm in several forms of protozoa. In majority of protozoa, the cytoplasm is differentiated into **ectoplasm** and **endoplasm**. The ectoplasm is gel like while the endoplasm is more voluminous and fluid. The membrane system of protozoa forms a more or less continuous network of canals and lacunae giving rise to the endoplasmic reticulum of the cell. Other structures in the cytoplasm include ribosome, golgi complexes or dictyosome, mitochondria, kinetosomes or blepharoplast, food vacuoles and contractile vacuole and nuclei. Every protozoan cell has at least one eukaryotic nucleus; some have multiple nuclei. In the ciliates, two dissimilar nuclei, one large (macro-nucleus) and one small (micro-nucleus) are present. The macronucleus controls the metabolic activities and regeneration processes; the micro-nucleus is concerned with reproductive activity.

The cytoplasm with its various structure is separated from the external environment by the cell envelope. It provides protection to the cell and controls the exchange of substances and is the site of perception of chemical and mechanical stimuli. The cell envelope sometimes contains a pellicle which in the simplest form (e.g., *Amoeba*) functions as cell membrane or the **plasmalemma**. The pellicle of a ciliate is thick, rigid and sculptured. There are other kinds of protective envelope produced by protozoa which are loose covering external to the pellicle like shell, test, lorica or cyst which vary considerably in their constituent materials.

Locomotion of protozoa is achieved by three types of organelles : pseudopodia, flagella and cilia; some may accomplish gliding movement by body **flexion**. A pseudopodium is a temporary projection of part of the cytoplasm (e.g., *Amoeba*) whereas the flagellum is filamentous extension of the cytoplasm. Cilia are fine, thread like extension from the cell and in addition to their locomotory function also aid in the ingestion of food.

Protozoa reproduce by a variety of asexual and sexual processes. Asexual reproduction occurs by binary fission or multiple fission or by budding. Binary fission is the most common method of reproduction in which the nucleus and other cell contents divide into two equal parts. Division occurs longitudinally in flagellate forms such as *Euglena*, transversely in ciliates like *Paramecium*. In multiple fission, a single mother cell divides to form many daughter cells. This type of division occurs commonly in the Foraminiferida, the Radiolaria and the Sporozoa. Budding is the formation of one or more smaller individuals from the parent organisms. It is restricted to those cases where mother cell remains sessile and releases one or more swarming daughter cells, as found in most of sessile ciliates. A characteristic method of reporduction, sporulation, is found in sporozoa. As in malaria parasite, there is a complicated life cycle involving more than one host.

In protozoa sexual reproduction also varies considerably. Sexual fusion of two gametes (syngamy or gametogamy) occurs in many groups of protozoa. Conjugation is found exclusively in Ciliophora. In this case, after exchange of nuclei, the conjugants dissociate from each other and each of them gives rise to its respective progeny by fusion or budding. The capacity of regenerating lost part is characteristic of all protozoa. When a protozoan is cut into two, the nucleated portion regenerates but the anucleated portion degenerates.

Protozoa occupy an important place in the food chain ot natural communities where free water is present. Of particular importance in the economy of many communities are the saprophytic and bacteria-feeding protozoa which are involved in the final decomposition level of food chains. Protozoa are also responsible for many human diseases of widespread occurrence. Malaria is a protozoan disease that has assumed alarming proportion throughout the world. The agents of **trypanosomiasis** (sleeping sickness), amoebic dysentry and giardic diarrohea are also quite common in many developing countries.

THE PROKARYOTIC CELL

The structure and functions of prokaryotic cells are described in Chapter-III. However, some more important characteristic features of prokaryotic cells are outlined here.

The prokaryotic cell is structurally less complex than the eukaryotic cell. They are rather small, about one twenty fifth of the volume of the eukaryotic cells. Compartmentation in prokaryotic cell is far less pronounced than the eucyte. There are no organelles like mitochondira and chloroplast, nor is the DNA surrounded by a nuclear membrane.

The nuclear region is visible as a network of fine threads on the cytoplasm which also contains large quantities of ribosomes. These ribosomes are smaller than the cytoplasmic ribosome of eukaryotes; they are of 70 S type. In the prokaryotes the cytoplasmic membrane is the site of respiratory and photosynthetic energy generation. Analogous functions in eukaryotes are localised in the membranes of mitochondria and chloroplast. The entire genetic information of the prokaryote is contained in a single thread of DNA, the bacterial chromosome.This DNA molecule exists as a circular strand in all bacteria. No histones are present. However, in many bacteria, extrachromosomal DNA (**plasmid**) have been identified.

The cells of prokaryotes (with very few exceptions, such as *Mycoplasma*), are surrounded by a cell wall. This contains a skeleton of **peptidoglycan** (also called murein) which is heteropolymer. Many prokaryotes are motile, moving by either swimming or gliding. The organs of locomotion are the bacterial flagella. These are much simpler in structure than eukaryotic flagella and contain a single kind of fibirl.

Reproduction in bacteria is generally by means of binary fission. After the appropriate increase in cell dimension septa appear, starting at the circumference and proceeding inwards, until eventually the two daughter cells separate. According to the number of cell division and divisions of plane, pairs of bacteria (Diplococci), chains (Streptococci), plates (Sarcina) and grape-like bunches (Staphylococci) can be distinguished. Multiplication by budding or sprouting is rare in prokaryotes. Prokaryotes are haploid. Cell division is preceded by doubling and replicating the bacterial chromosome. A diploid phase is therefore limited to a very short stage of the cell division cycle.

PROKARYOTIC MICROBES

Prokaryotic cells are divided into two major groups: **cyanobacteria** and **bacteria**.

CYANOBACTERIA : Cyanobacteria (Cyanophyta: Myxophyceae), formerly called blue green algae, are prokaryotes that perform oxygen-evolving photosynthesis in a manner similar to eukaryotic algae and plants.

The morphology is basically that of unicells or aggregation of these to form colonies and filaments. The cell walls are indistinct but often there is a prominent sheath of mucilage around the cell or filament. Under the microscope they show bluish-green colour but distinct **chromatophers** are absent, the pigments being diffused throughout the cells. The protoplasts consist of central region, the **nucleoplasm** containing the nuclear material without any binding membrane and an outer chromatoplasm in which the pigments are dispersed. Chlorophyll *a*, B - carotene, phycobilin, phycocyanin and phycoerythrin are the common pigments of Cyanophycean cells. Electron microscope shows a series of lamellae (**thylakoids**) mainly in the peripheral regions of the cell but also penetrating into the centre. Although these are not bounded by any enclosing membranes they are certainly the sites of photosynthesis. The nuclear material appears as threads. Under the light microscope the cell is seen as a bipartite structure (inner and outer investment) but clearly there are three regions in the inner investment. These are, however, not unit membrane. Several cross walls can be seen within one cell which is a characteristic feature of many filamentous cyanobacteria.

Fine pores of **plasmodesmata** also penetrate the cross walls and in some species (Stigonematinales) **central pit connection between the cells are seen.**

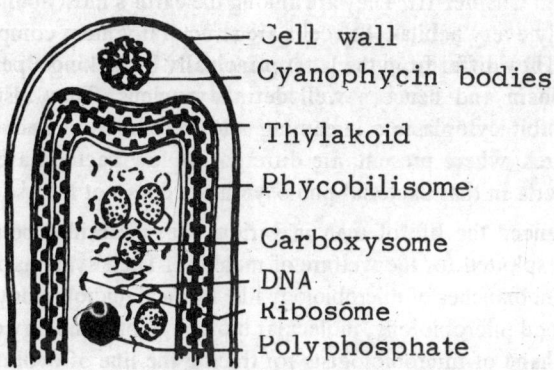

Cell wall
Cyanophycin bodies
Thylakoid
Phycobilisome
Carboxysome
DNA
Ribosome
Polyphosphate

Fig. 2.1 Longitudinal section through cells of cyanobacteria.

Vegetative growth is the only known method of increase in cell number of many cyanobacteria especially of coccoid series (Chroococcales). Few genera form exospores (e.g., *Chamaesiphon*). Some filamentous genera differentiate small segments of the filaments (**hormogonia**) which are more actively motile and form a vegetative means of propagation. These are sometimes surrounded by a thicker wall and are then incapable of movement, the so called **hormocysts**. Both types eventually germinate by outgrowth of the filament. Enlarged cells, with dense contents, sometimes with inflated walls occur singly or in rows in *Anabaena, Cylindrospermum* etc. and act as **akinetes**. These resting spores also germinate to form a new filament. Sometimes an apparently empty, slightly enlarged cells, the **heterocysts**, are often associated with akinetes which do germinate to form filaments.

Cyanobacteria share many characters with algae, chiefly their ability to photosynthesize via chlorophyll *a* with the liberation of oxygen. At the same time they possess many features of bacteria such as :

(i) absence of nuclear membrane and membrane bound chloroplast of endoplasmic reticulum, mitochondria and golgi apparatus.

(ii) presence of peptidoglycan and diaminopimelic acid in the cell wall.

(iii) infection with viruses that resemble bacteriophages.

(iv) overall morphological simplicity of bacteria.

These features, showing relatedness with bacteria, led many microbiologists and even phycologists to consider blue green algae to be bacteria.

· Cyanobacteria have captured the attention of microbiologists in recent decades mainly for two notable features :

(i) ability of some species to transform molecular nitrogen into a form usable by plants, thereby introducing the essential nutrients into the food chain; as symbiotic partner of higher plants (e.g., *Anabaena* with *Azolla* or *Nostoc* with *Anthoceros* etc. they contribute significantly in nitrogen fixation of soil.

(ii) ability of some species to produce very potent toxin (e.g., by *Microcystis, Anabaena* and *Aphanizomenon* etc.) which have been responsible for the death of cattle.

BACTERIA

All prokaryotes that are not cyanobacteria are classified as bacteria. Detailed structure and classification of bacteria will be presented in Chapter-III. They are among the earth's most abundant and diverse organisms which are found in virtually every habitat. The cells are structurally more complex than viruses but less so than those of eukaryotes. They differ from the latter principally by lacking a perforated nuclear membrane surrounding the **nucleoplasm** and hence a well defined nucleus. They also lack organelles such as mitochondria, do not exhibit cytoplasmic streaming and the structure and composition of cell-walls, flagella and chromatophores, where present, are different. A few bacteria are photosynthetic but these differ from the cyanobacteria in that bacterial photosynthesis does not release oxygen.

Bacteria have influenced the life of man in various ways. Besides being pathogenic to man and animal, these are largely exploited for the welfare of mankind. Their systematic studies have fostered the growth of many independent branches of microbiology like medical microbiology, agricultural microbiology, industrial microbiology, food microbiology, molecular biology, biotechnology etc. They are the most ideal experimental tools in the hand of microbiologists for tracing the line of evolution on earth.

The above illustrations of microorganisms based on recent advances in morphological and biochemical techniques, reveal fundamental resemblances and differences of various microorganisms and offer a sound basis for division into major group. A schematic classification of microorganism is depicted in Table-2.3.

Table 2.3

CLASSIFICATION OF MICRO-ORGANISMS

A.	Organisation subcellular	VIRUSES
AA.	Thallus unicellular, multicellular or plasmodial	
B.	Nucleoplasm not bounded by a membrane	PROKARYOTA
C.	Chlorophyll absent, or if present, type different from that of plants	BACTERIA
CC.	Chlorophyll present together with characteristic blue-green pigments not located in discrete plastids	CYANOBACTERIA
BB.	Cell or plasmodia containing one or more discrete membrane bounded nuclei	EUKARYOTA
D.	Cells of vegetative thallus possessing cell wall	
E.	Chlorophyll present and located in discrete chloroplasts	ALGAE (excluding the blue-green algae)
EE.	Chlorophyll absent	FUNGI
DD.	Cells of vegetative thallus lacking true cell walls	
F.	Thallus unicellular, remaining so	PROTOZOA
FF	Thallus unicellular at first, becoming a plasmodium or pseudoplasmodium and, eventually forming a fructification .	SLIME MOULDS (Myxomycetes or Mycetozoa)

SUGGESTED READINGS

Ainsworth, G.C. (1975). Ainsworth and Bisby's Dictionary of the fungi, 6th Edn. Commonwealth Mycological Institute, Kew, Surrey.

Ainsworth, G.C. and Sussman, A.S. (ed.) (1965) The Fungi. Vol. I. The Fungal Cell (1966) Vol.II. The Fungal Organism (1968), Vol.III. The Fungal Population (1973), with F.K. Sparrow, Vol. IV A. Taxonomic Review with keys: Ascomycetes and Fungi Imperfecti. Vol. IV B. Basidiomycetes and Lower Fungi. Academic Press, New York.

Alberts, B., Bray, D., Lewis, J., Raf, M., Robert, K. and Watson, J.D. (1983). Molecular Biology of the Cell. Garland Publishing, New York.

Alexopoulus, C.J. (1962). Introductory Mycology, 2nd Edn. Wiley, New York.

Baker, J.R. (1969). Parasitic Protozoa. Hutchinson, London.

Burnett, J.H. (1976) Fundamentals of mycology 2nd eds. Edward Arnold, London.

Cavelier - Smith, T. (1981). The origin and early evolution of the eukaryotic cell. Sym. Soc. Gen. Microbiol. **32**: 33.

Corless, J.O. (1973). The fine structure of algal cells. Pergamon Press, Oxford.

Dodge, J.D. (1973). The fine structure of algal cells. Academic Press, London and New York.

Dworkin, M. (1966). Biology of Myxobacteria. Ann. Rev. Microbiol. **20**: 75-106.

Echlin, p. (1966). The Blue green Algae. Scient. Am. **214**: 74-83.

Fogg, G.E., Stewart, W.D., Fay, P. and Walsby, A.E. (1973). The Blue green Algae. Academic Press, London and New York.

Grell, K.G. (1973). Protozoology. Springer - Verlag, Heidelberg.

Holm - Hansen, O. (1968). Ecology, Physiology and Biochemistry of the Blue-green algae. Ann. Rev. Microbiol. **22**: 47-70.

Kelly, D.P. and Carr, N.G. (eds.) (1984). The Microbe. Symp. Soc. Gen. Microbiol.**36**: 11.

Lang, N.J. (1968). The fine structure of Blue-green algae. Ann. Rev. Microbiol. **20**: 75-106.

MacKinnon, D.L. and Hawes, R.S.J. (1961). An introduction to the study of Protozoa.. Oxford University Press, London.

Pitelka, D.R. (1963). Electron microscopic structure of Protozoa. Pergamon Press, Oxford.

Shively, J.M. (1974). Inclusion bodies of Prokaryotes. Ann. Rev. Microbiol. **28**: 167-187.

Stainer, R.Y., Rogers, H.J. and Ward, J.B. (1978). Relations between Structure and Functions in the Prokaryotic cell. 28th Symp. Soc. Gen. Microbiol. Cambridge University Press, Cambridge.

3

THE PROKARYOTES : STRUCTURE AND FUNCTION

The prokaryotes include the groups of Schizomycetes (bacteria in the wider sense) and Schizophycetes (cyanobacteria). Traditionally the latter and the Actinomycetes of Part 17 of Bergey's Manual are studied with algae and fungi respectively. For this reason detailed account of structure and function of Bacteria is presented here.

BACTERIA

Bacteria include a great variety of microorganisms of different configurations and dimensions. Most bacterial species are about 1 μm wide and 5 μm long. Many pseudomonas have a diameter of 0.4-0.7 μm and a length of 2-3 μm. The diameter of micrococci is only 0.5 μm. Owing to their minute size, bacteria can be examined under the light microscope only with objectives of high resolving power. Most bacteria are transparent and have a refractive index similar to that of the aqueous fluids in which they are suspended. In order to render them easily visible stained preparation are often used (See Box. 1). More recently phase contrast microscopy has been used for observing living cells.

MORPHOLOGY

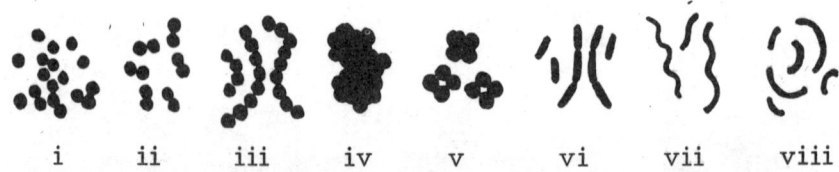

| i | ii | iii | iv | v | vi | vii | viii |

Fig. 3.1 Different morphological forms of bacteria.

i. Micrococci ii. Diplococci iii. Streptococci iv. Staphylococci v. Sarcina vi. Rod-shaped bacteria vii. Spirilla viii. Vibrios.

The shape of nearly all bacteria, with few exceptions, can be derived from spheres, cylinders and curved cylinders. The basic forms, therefore, are **cocci, straight rod** and **curved rods** (Fig. 3.1). The morphology of different cells in a population are not always identical; variation in size and form of the individual cell occurs with age and growth conditions. This variation or **pleomorphism** is particularly characteristic of some organisms. In addition, under certain conditions many bacteria may change their form and structure to produce involution and **L-forms**.

(BOX 1)

In 1884 Christian Gram developed a method for differentiation of bacteria in the tissue sections. Based on their reactions to stain, bacteria are grouped into Gram-positive and Gram-negative forms.

The organisms are stained with a basic dye, **crystal violet**, at slightly alkaline pH, followed by mordanting usually with iodine in a solution of potassium iodide or picric acid. The crystal violet combines with the iodine to form a complex. Subsequent washing of the stained preparation with a neutral (organic) solvent (usually ethanol or acetone) causes the crystal violet-iodine complex to be eluted from the Gram-negative species whilst the Gram-positive organisms retain the stain. Counter staining with another dye of contrasting colour (eg., dilute fuchsin, safranine) renders the difference more striking. The Gram-positive bacteria appear as blue film whereas Gram-negative look red.

Spheres

Most cocci (= berry L.) approximate the true sphere, though deviations do occur. *Staphylococcus aureus* and *Streptococcus pyogenes* are spherical whereas *Streptococcus faecalis* is ellipsoidal and *Neisseria gonorrhoeae* is kidney-shaped. Similarly *Streptococcus pneumoniae* shows division in one plane giving rise to pairs of organisms whereas *Streptococcus pyogenes* forms chain. Regular division in two planes at right angle to each other results in tetrad (e.g., *Micrococcus roseus)* and division at three planes to cubical packet of eight or more is characteristic of genus *Sarcina.* Irregular division forming grape-like clusters is special feature of *Staphylococcus aureus.* The colony so produced as a result of cell division is to taxonomic significance for bacterial identity.

Straight rods

Organisms of this morphological form were formerly referred to as bacillus (= stick L.). These are straight or very slightly curved cyliniders, the ratio of length to diameter varying cosiderably. Some typical representatives of this form are *Serratia marcescens, Nitrosomonas europaea, Haemophilus influenzae, Bacillus anthracis* etc. (Fig. 3.2). End walls of rods are usually convex (e.g., *S. marcescens*) but may be flat

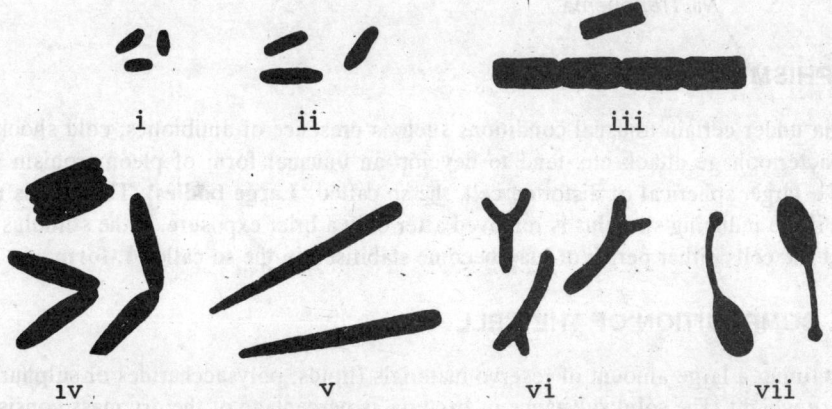

Fig. 3.2. Various rod-shaped bacteria

i. *Serratia marcescens* ii. *Escherichia coli* iii. *Bacillus anthracis* iv. *Corynebacterium* sp. v. *Fusobacterium fusiformis* vi. *Bifidobacterium* vii. *Hyphomicrobium* (e.g., *B. anthracis*) or club-shaped at one end (e.g., *Corynebacterium* sp.) or tapering (*Fusobacterium fusiformis*). In another related bacteria the rods show true branching either with development into filament (e.g., *Nocardia*) or without (*Bifidobacterium*). Some produce an extensive mycelium (*Streptomyces*) which fragments to form spores. Rods usually do not reproduce by binary fission or fragmentation but by producing buds on appendages. Among the Myxobacteria, some organisms exhibit simple differentiation, the vegetative rods rounding off to form **myxospore**; others form large structure called fruiting bodies possessing stalk cells, sporangial wall cells and myxospores.

Curved rods

Vibrio cholerae is the best known example of a curved rod. *Bdellovibrio* also exists as a small curved rod in its parasitic phase. Other representatives of this form are *Spirillum, Spirochaeta, Cristispira, Borrelia, Treponema* etc. which vary in their size but have typical curved orientation. (Fig. 3.3).

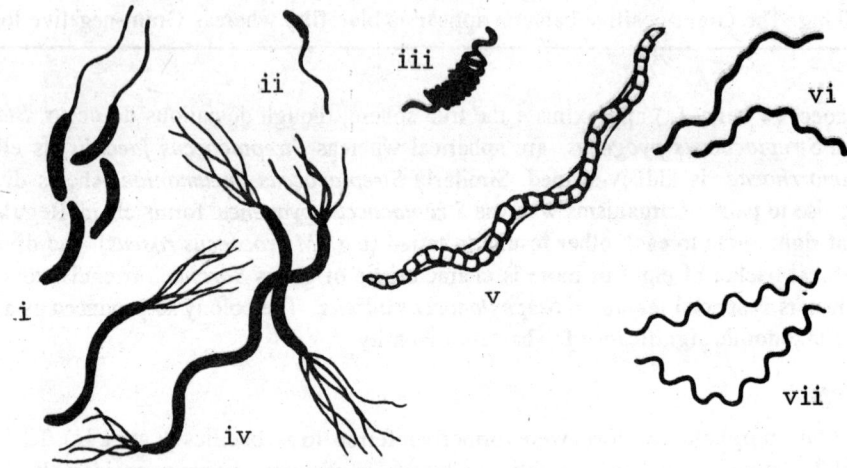

Fig. 3.3 Various curved rod bacteria

i. *Vibrio cholerae* ii. & iii. *Bdellvibrio* iv. *Spirillum* v. *Cristispira* vi. *Borrelia* vii. *Treponema*

PLEOMORPHISM

Many bacteria under certain unusual conditions such as presence of antibiotics, cold shock, presence of antiserum, bacteriophage attack etc. tend to develop an unusual form of pleomorphism involving the formation of a large, spherical or distorted cell, the so called **'Large bodies'**. These cells revert to their normal form if the inducing stimulus is removed after only a brief exposure. If the stimulus persists for a longer period the cells either perish or may become stabilised in the so called **L-form**.

MATERIAL COMPOSITION OF THE CELL

Bacteria containing a large amount of reserve materials (lipids, polysaccharides or sulphur) have higher percentage dry weight. The solid substance of bacteria as percentage of the dry mass consists-of proteins 50; cell-wall, 10-20; RNA, 10-20; DNA, 3-4; lipid about 10. The percentage bioelements represented in the composition of bacteria are: carbon, 50; oxygen, 20; nitrogen, 14; hydrogen, 8; phosphorus, 3; sulphur, 1; potasium, 1, calcium, 0.5; magnessium, 0.4 and iron, 0.2.

ULTRA-STRUCTURE OF BACTERIAL CELL

The Gram-stain used in bacteriology reveals little of the internal structure of the organism but the gross morphology. For this reason many special staining techniques such as **Feulgen method** for nuclear bodies, the **Malachite green method** for endospores, the **Fontana method** for flagella etc. have been devised to reveal details of particular structure. The use of **'Phase-contrast'** and **'Dark-ground'** microscopy has led to the acquisition of further information but by far the most powerful tool for studying bacterial anatomy is the Electron Microscope in which magnification upto 2,50,000 dimension can be achieved.

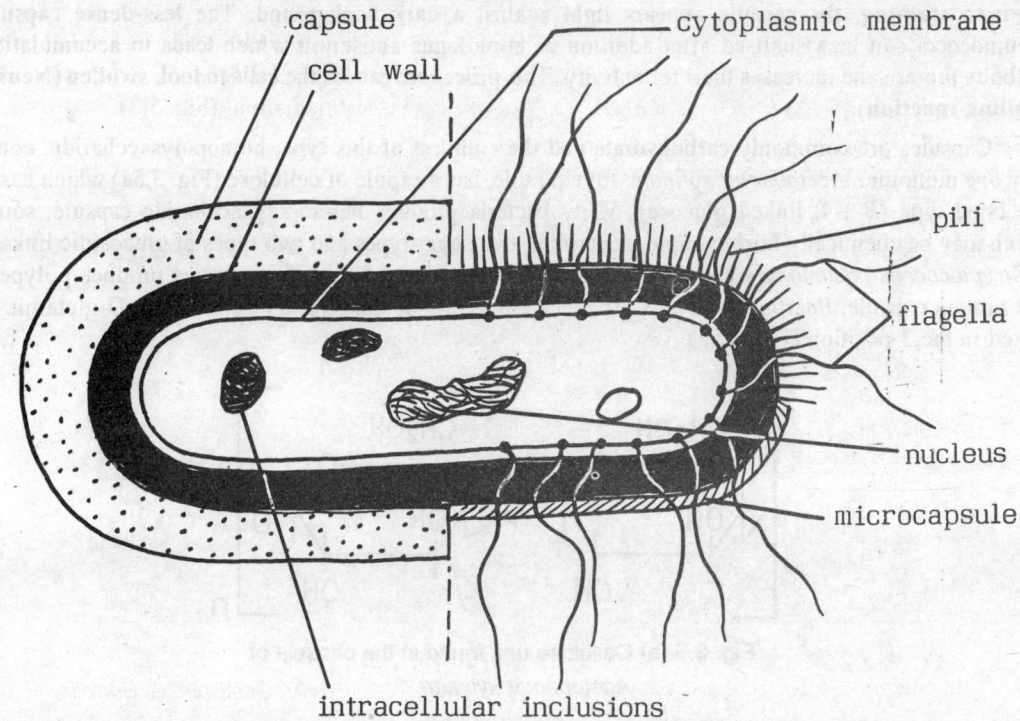

Fig. 3.4 A typical bacterial cell. Right to the dotted line represents the non-capsulated, flagellate rod.

The principal structures of a typical bacterial cell are shown in Fig. 3.4. In general, many bacteria are surrounded by a **capsule** beneath which is a **cell wall** giving shape to the cell. Some have filamentous appendages **fimbriae or pili**; some have one or many **flagella**. Beneath the cell wall is a fine **plasma membrane**. Folded invagination of plasma membrane, the **mesosome**, are present in many bacteria. The **nucleoid** is in the form of a highly folded DNA ring without a nuclear envelope and the cytoplasm contains 70 S ribosome, internal membrane system, storage granules, vacuole chromatophore (in photo-synthetic bacteria). The details of these structures are described below.

CAPSULES, SLIME LAYERS AND SHEATH

Many bacteria accumulate layers of water rich materials on their outer surface. According to the amount of material produced and the degree of its association with the cell, one of the three terms may be applied but it is difficult to define exactly the borderline between them. If the material is closely adherent to the cell and detectable only with difficulty, it is known as **microcapsule**; if equally sharply defined but extensive and readily visible using appropriate technique it is called a **capsule**; if copious in quantity and only relatively loosely associated with the cell it is referred to as **slime layer**. These envelopes are usually not essential for bacterial life, but the possession of capsules makes some pathogenic bacteria resistant to **phagocytosis** and enhances their virulence in experimental animals.

Capsule

After addition of dyes that do not penetrate the capsular material such as chinese ink, nigrosin, or congo red, it is easy to demonstrate the presence of capsule with the light microscope. In this technique, called **negative staining**, the capsule appears light against a dark background. The less-dense capsule of pneumococci can be visualised after addition of homologus antiserum which leads to accumulation of antibody protein and increases their refractivity. This procedure causes the cells to look swollen (**Neufeld's swelling reaction**).

Capsules are commonly carbohydrate and the simplest of this type, homopolysaccharide, contains only one monomer. *Acetobacter xylinum.* for example, has a capule of cellulose (Fig. 3.5a) which has only one bond type (*B*-1-4, linked glucose). Many bacteria produce heteropolysaccharide capsule, some of which may be chemically fairly simple, containing two sugar types and two types of glucosidic linkage as in *Streptococcus pneumoniae* and *Streptococcus* species (Fig. 3.5b). A few species produce polypeptide and protein capsule. *Bacillus anthracis* produces a polymer of the unusual amino acid, D-glutamic acid; linked in the 3 position (Fig. 3.5c).

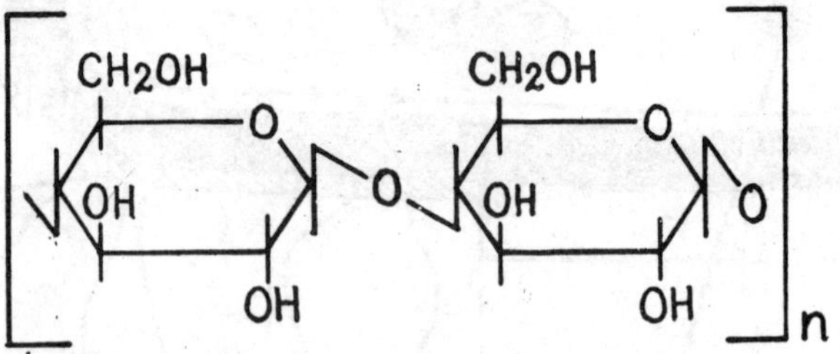

Fig. 3.5 (a) Cellulose unit found in the capsule of
Acetobacter xylinum.

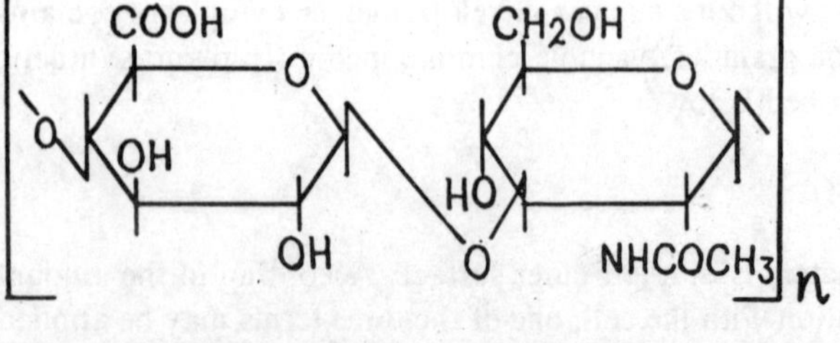

Fig. 3.5 (b) Hyaluronic acid capsule of *Streptococcus.*

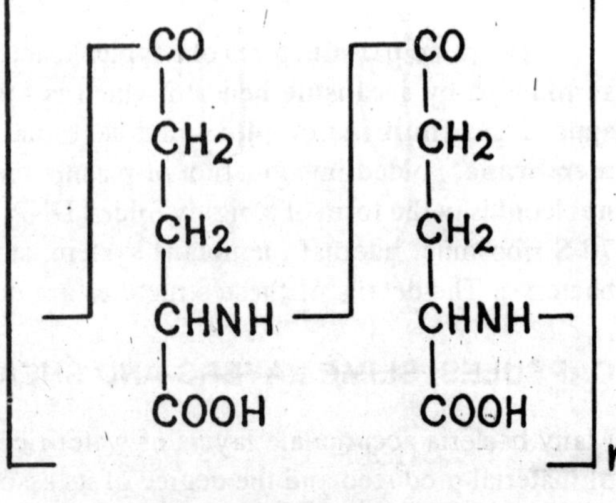

Fig. 3.5 (c) Poly D-glutamic acid capsule in
Bacillus anthracis

Capsules are not indispensible for cell structure because (i) they are not synthesised under all environmental conditions (ii) the mutants exist which have lost the ability to produce capsule and (iii) cells from which they have been removed by enzymic digestion remain viable. The presence of capsule, however, is often associated with the virulence of pathogenic organism. When capsulated organisms enter the animal body they are able to resist phagocytosis by the white cells of the blood i.e., they are not readily engulfed by the scavanger cell, or if they do become engulfed they resist digestion by the enzyme and continue to multiply. Loss of capsule, not only leads to loss of virulence, but is often accompanied by other changes such as alteration of the colony structure. It is also suggested that capsule may act as ion exchangers, absorbing nutrients and subsequently releasing them for use by the cell. It is also involved in the prevention of bacteriophage attack and of dessication.

Slimes

Much of the capsular materials may be excreted into the medium as slime. In some cases the entire capsular material can be removed from the cell surface by shaking or homogenising the bacterial suspension and can be recovered from the medium as slime. A more extensive slime formation occurs with many microorganisms when the medium is supplemented with saccharose. A well known example of this is *Leuconostoc mesenteroides*, a heterofermentative lactic acid bacterium known in sugar refining plants as 'frog-spawn bacterium' which rapidly converts a solution containing cane sugar to a stiff jelly consisting of dextran.

Sheath

Some filamentous bacteria form tubular envelopes described as sheath (*Sphaerotilus natans, Leptothrix ochracea*). These sheath consist of a heteropolysaccharide containing glucose, glucuronic acid, galactose and fucose. *Acetobacter aceti* var. *xylinum* secretes cellulose which forms the **'mycodermaceti'** a leather like skin that surrounds the cells. The cells of *Sarcina ventriculi* are bound by the excreted cellulose in regular association. The cellulose in these cases serves as a kind of mortar and is quite different from capsule, both structurally and functionally.

FLAGELLA

Bacteria can achieve motility in a number of ways. In most of the actively motile swimming bacteria motility is brought about by the rotation of flagella (singular, flagellum = whip L.). Flagella are unbranched,

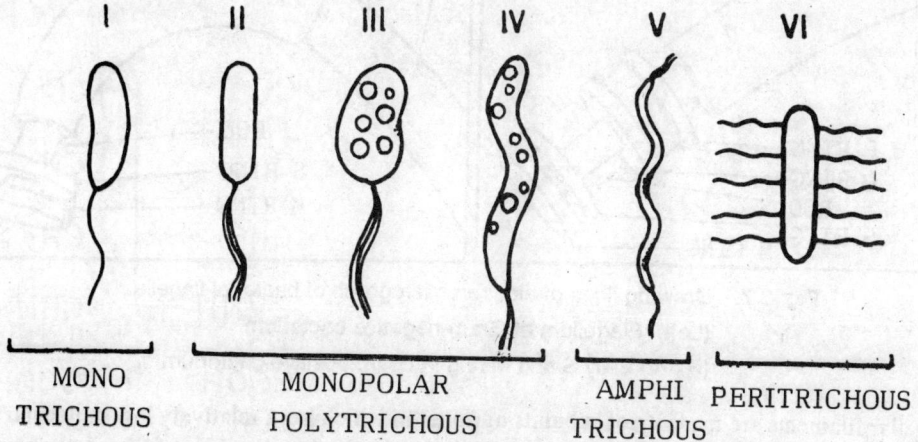

| I | II | III | IV | V | VI |

| MONO TRICHOUS | MONOPOLAR POLYTRICHOUS | AMPHI TRICHOUS | PERITRICHOUS |

Fig. 3.6 Some important types of bacterial flagellation

i. *Virbio* ii. *Pseudomonas* iii. *Chromatium* iv. *Thiospirillum*
v. *Spirillum* vi. *Proteus*.

helical filaments of uniform thickness, about 20 nm throughout their length. Their size is too small to be seen under the light microscope but considerable thickening by means of metal salt depositon or special staining technique, such as those using fluorescent dyes, render them visible.

The ways in which flagella are arranged on the bacterial cell are highly specific for different motile bacteria and hence are of taxonomic value. A single bacterium may possess from one to hundred flagella (Fig. 3.6). Amongst the bacteria with monopolar flagellation some have only single flagellum (*Vibrio metchnikovii*), while many (*Pseudomonas, Chromatium, Thiospirillum*) consist of bundle of 2 to 50 single units (**polytrichous**). Monopolar polytrichous flagellation is also referred to as **lophotrichous**. Bipolar polytrichous flagellation is called as **amphitrichous** (e.g., *Spirillum*). In pertitrichously flagellated bacteria (*Escherichia, Proteus, Clostridium*) the flagella are inserted in the lateral walls or over the entire surface.

The flagellum has three parts : a basal structure, a hooklike structure and a long filament outside the cell wall. It has been obvious for sometime that flagella probably originate inside the cell wall, because organisms from which the cell wall has been removed (i.e., protoplast) retain their flagella. Electron micrographs show the proximal ends well within the cytoplasm. Until recently it has been thought that flagella originate in a vesicle in the cytoplasm but the current view is that the basal structure is partly in the cytoplasmic membrane and partly in the wall. The basal structure is complex and different in Gram-positive and Gram-negative cells (Fig. 3.7). In *E. coli* (a Gram-negative bacterium) four rings (M = membrane; S = super membrane; P = peptidoglycan; L = lipopolysaccharide) are mounted on a rod attached to the proximal end of the hook. The M ring binds to the preparation of the inner cytoplasmic membrane and the L ring binds to the preparation of the outer lipopolysaccharide membrane of the cell wall. In Gram-positive bacteria (*Bacillus subtilis*), the L and P-rings are not present. As in Gram-negative bacteria in this case too, M ring binds to the cytoplasmic membrane and the S ring is supposed to be attached to the inside of the thick peptidoglycan layer. Outside the wall is a short length of filament, the **hook** (about 0.05 μm long). It is somewhat thicker than the rest part of the flagellum and is different in its chemistry. The remainder and the largest part of the flagellum is the filament which, in electron micrograph, looks bead-like subunits and is usually helically arranged.

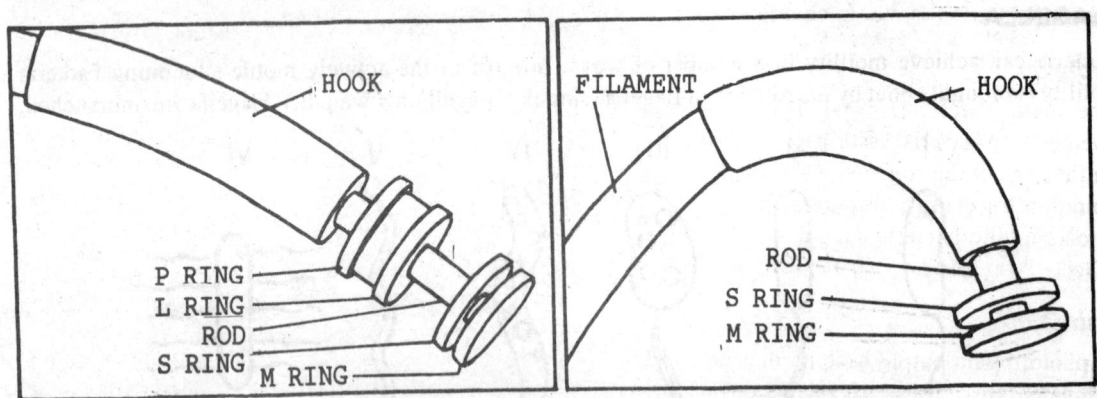

Fig. 3.7.   Drawing lines of electron micrograph of bacterial flagella.

(Left) Flagellum of Gram-negative bacterium

(Right) Only S and M ring in Gram-positive bacterium

Flagellar filaments are made up of subunits of **flagellin** which has a relatively low molecular weight. These subunits are arranged in helical order around an axial cylinder. Analysis of the components show that more than 98% of the total weight is protein which characteristically has a high acidic amino acid content, few aromatic amino acid residues and no cysteine. The flagellar proteins constitute the diagnostically important **H-antigen**.

Flagella are responsible for movement. The speed attained vary considerably, for example, *Bacillus megaterium* can move 1.6 mm/min and *Vibrio cholerae* can move 12 mm/min. Polar flagellation shows greater mobility than peritrichously arranged flagella (only 12 μm/min). The means by which flagella cause bacterial movement were uncertain till recently. It was believed that by sequential alteration of the configuration of flagellin molecule, a wave was propagated in the flagellum. This triggers the whole helical structure rotate. In most cases of polar flagellation, the flagellin acts like a ship's screw and pushes the bacterial cell through the medium. The basal structure is analogous to an electric motor with S and M rings acting as the 'stator' and 'rotor' respectively. this rotation motor rotates around a fictitious axis of the screws. However, the molecular mechanism of the flagellar 'rotary' motor is not yet known. The flagellar rotation enables the bacterial body to rotate in the opposite direction. Flagella can also change the orientation of their rotation either spontaneously or in response to external stimuli.

In spirochaetes a less common type of locomotion is exhibited. These helically shaped organisms move due to **flexions** of the cell caused by axial fibres running along the long axis of the cell but within the flexible outer envelope. The gliding bacteria (Myxobacteriales) also have flexible cell wall and move as a result of its blending. These organisms move comparatively slower than the flagellate bacteria.

RESPONSES TO EXTERNAL STIMULI

Chemotaxis

Motile bacteria are able to move in definite directions i.e., they exhibit **taxis**. When they react to chemical stimuli, they accumulate in some areas or retreat from others. Such stimulus-response behaviour is called **chemotaxis**. Peritrichously flagellate bacteria have two kinds of motile behaviour, straight-line swimming and tumbling, which interrupts the straight movement and causes reorientation. If the bacteria are placed in a concentration gradient of an attractant, the linear swimming movement occurs which lasts for many seconds in the direction of the optimal concentration of the attractant. The tumbling movement leads to a completely random selection of the new swimming direction. The linear swimming motion results in accumulation of the organisms in region of the optimal substrate concentration. The sensing of and response to stimuli is due to **chemoreceptors**.

Aerotaxis

Some motile bacteria reveal their metabolic capacities relative to oxygen or air by their aerotactic movements. In bacterial suspensions placed between slide and coverslip aerophilic microorganisms accumulate near the edge of the converslip and in the vicinity of air bubble, demonstrating their dependence on aerobic respiration for energy. Strictly anaerobic bacteria, on the other hand, tend to collect in the centre while microaerophilic bacteria such as some pseudomonads and spirilla keep a certain distance from the air surface.

Phototaxis

The phototrophic purple bacteria depend on light for their energy supply. They, therefore, accumulate in illuminated area. These organisms do not leave the illuminated area, once they have entered it, in the course of their random movement. On entering the dark zones, abrupt reversal of flagellar motion propels them back into the light zone. This reversal is so sudden that this response has been called **'phobataxis'** (shock reaction). Even slight difference in light intensity between area of illumination can sometimes evoke this reaction.

Magnetotaxis

Recently a number of bacteria isolated from the surface layers of sediments in fresh water ponds have been found to orientate themselves in a magnetic field and swim in the direction of the field lines. They contain unusual amount of iron (0.4% of their dry weight) as ferromagnetic iron oxide in the form of **grana** (magnetosome) which are localised close to the area of flagellar insertion.

PILI (FIMBRIAE)

Pili or fimbriae are very fine, straight filamentous appendages much smaller than flagella which are usually found in Gram-negative bacteria (*Corynebacterium renale*, a Gram-positive bacterium being the only exception). They measure less than 10 μm in diameter and appoximately one μm long. The number per cell varies from one to 400 and there may be one or more morphological types present. These can be seen only by electron microscopy.

Pili are like flagella in that they are cytoplasmic extrusion through the cell wall. They originate from the cytoplasm and penetrate through the peptidoglycan layers of the cell wall. They consist of protein subunits, **fibrilin**, having molecular weight about 17000 as compared to about 40,000 for the flagellar protein **flagellin**. It consists of about 163 amino acids which occur mostly in the L-form with a large amount of hydrocarbon side chains.

Based on their functions at least ten types of pili have been recognised so far. **Sex pili** (or F - pili) have been studied in great detail which are required for transfer of DNA between donor and recipient cells during conjugation. These are specialised filamentous structures which are determined by sex factors (plasmid). These sex factors control the antigenic features of the sex pilus and its capacity to absorb bacteriophages. The sex pilus appears to have an axial hole 25-30 A in diameter. Electron micrographs reveal the presence of a terminal knob 150-800 A in diameter. In *E. coli* two types of sex pili can be distinguished; **F-pili** and **I-pili**. F-pili are determined by F or F-like R-factors and the I-pili by Col I or I like R-factors. Immunological studies provide evidence that tip of the pilus senses the female (F$^-$) cell and constitutes a specific site for attachment.

Besides sex pili, other types of pili cause bacteria to adhere to natural substrate and other foreign materials such as red blood cells. Piliated strains of *E. coli* may be recognised by their ability to haemagglutinate red blood cells. It has also been seen that pili can adhere to the epithelial cells and contribute significantly to the adhesion of *Neisseria gonorrhoeae* to the urethra wall. Pili are also found to effect the metabolic activity of bacterial cells. In culture of *E. coli* K 12 the respiratory activity of Fim$^+$ cell is considerably greater than that of Fim$^-$ cells. This supports the view that fimbriae function as aggregation and increase the oxygen supply by forming pellicles on the culture medium.

THE CELL WALL

Beneath the capsule or slime and external to the delicate cytoplasmic membrane is the cell wall rigid structure which gives shape and structural integrity to the bacterial cell. The thickness of cell wall ranges from 10 to 25 nm that constitute a significant portion of total dry weight of the cell. Bacterial cell walls seem to be essential for bacterial growth and division. Cells whose walls have been selectively removed i.e., **protoplast**, are incapable of normal growth and division.

Pure prepartion of bacterial cell walls can be obtained by mechanical disintegration of the cells such as exposure to sonic or ultrasonic treatment or exposure to extremly higher pressure with sudden release and subsequent separation of cell wall fragments by differential centrifugation. Autolysis, osmotic lysis and heat shock have also been used for cell wall preparations. Chemical and enzymic methods are not used now because these alter the chemical composition.

The principal structural polymer of both Gram-positive and Gram-negative bacteria is **peptidoglycan** (mucopeptide murein). Other interesting chemicals are diaminopimelic acid, muramic acid and teichoic acid. Typical composition of cell walls of Gram-positive and Gram-negative bacteria is shown in

Table 3.1

CHEMICAL COMPOSITION OF CELL-WALL OF
GRAM-POSITIVE AND GRAM-NEGATIVE BACTERIA

	Gram-Positive	Gram-Negative
Monomers		
Amino acids	~ 4	~ 18
Amino sugars	N-acetylglucosamine and acetylmuramic acid	Same
Lipids	< 2%	< 20%
Polymers		
Peptidoglycan	+	+
Teichoic acid or teichuronic acid	+	-
Lipopolysaccharide	-	+
Lipoprotein	-	+

The Basic Skeleton of Cell Wall

The supporting skeleton of the bacterial cell wall consists of a regular polymer, the peptidoglycan of murein. This heteropolymer is made up of chains of alternating molecule of **N-acetyl glucosamine** (GlcNAc) and its lactic acid ether, **N-acetyl muramic acid** (MurNAC) linked by 1, 4-B-glycosidic bonds. The muramic acid unit in the chain have short peptide attached to their lactyl residue by peptide bonds. The typical amino acids of these peptides are L-alanine, D-glutamic acid, m-diaminopimelic acid or L-lysine and D-alanine. The diamino acid, m-(or LL)-diaminopimelic acid and L-lysine play an important role in the formation of an intramolecular network because both of the amino groups can take part in peptide bond formation. Thus they can contact two of the straight heteropolymer chains (Fig. 3.8). Diaminopimelic acid or lysin may be replaced in some cases by ornithine or diaminobutyric acid. The heteropolymer chains thus connected via their peptide side chains form a sac like molecule, the **'murein sacculus'**.

The murein sacculus functions as the supporting skeleton of the cell wall and is penetrated and surrounded by a number of other substances. This, however, differs considerably in Gram-positive and Gram-negative bacteria.

The Cell Wall of Gram-Positive Bacteria

In Gram-positive bacteria the murein network represents 30-70% of the dry weight of the cell wall and consists of about 40 layers. In *Staphylococcus aureus* the tetrapeptide side chains of the muramic acid are connected by interpeptide chains (such as pentaglycine). Polysaccharide, if present in the cell wall of Gram-positive organisms are covalently bound whilst the protein content is very minor. The presence of teichoic acid is characteristic of Gram-positive bacteria. Teichoic acids in many instances represent a major component of the wall accounting for upto 10% of dry weight of cells. These are anionic polymers which always contain either glycerol or ribitol usually combined with sugar and D-alanine (Fig. 3.9). One molecule of techoic acid is covalently linked through a phosphodiester bridge to one muramic acid residue in each mucopeptide glycan chain. In some organisms according to environmental condition teichuronic acid is synthesised alternatively to teichoic acid. Teichoic acid may be involved in the regulation of autolytic enzymes, in the binding of cations and in the maintenance of the correct level of magnessium ions for membrane stability.

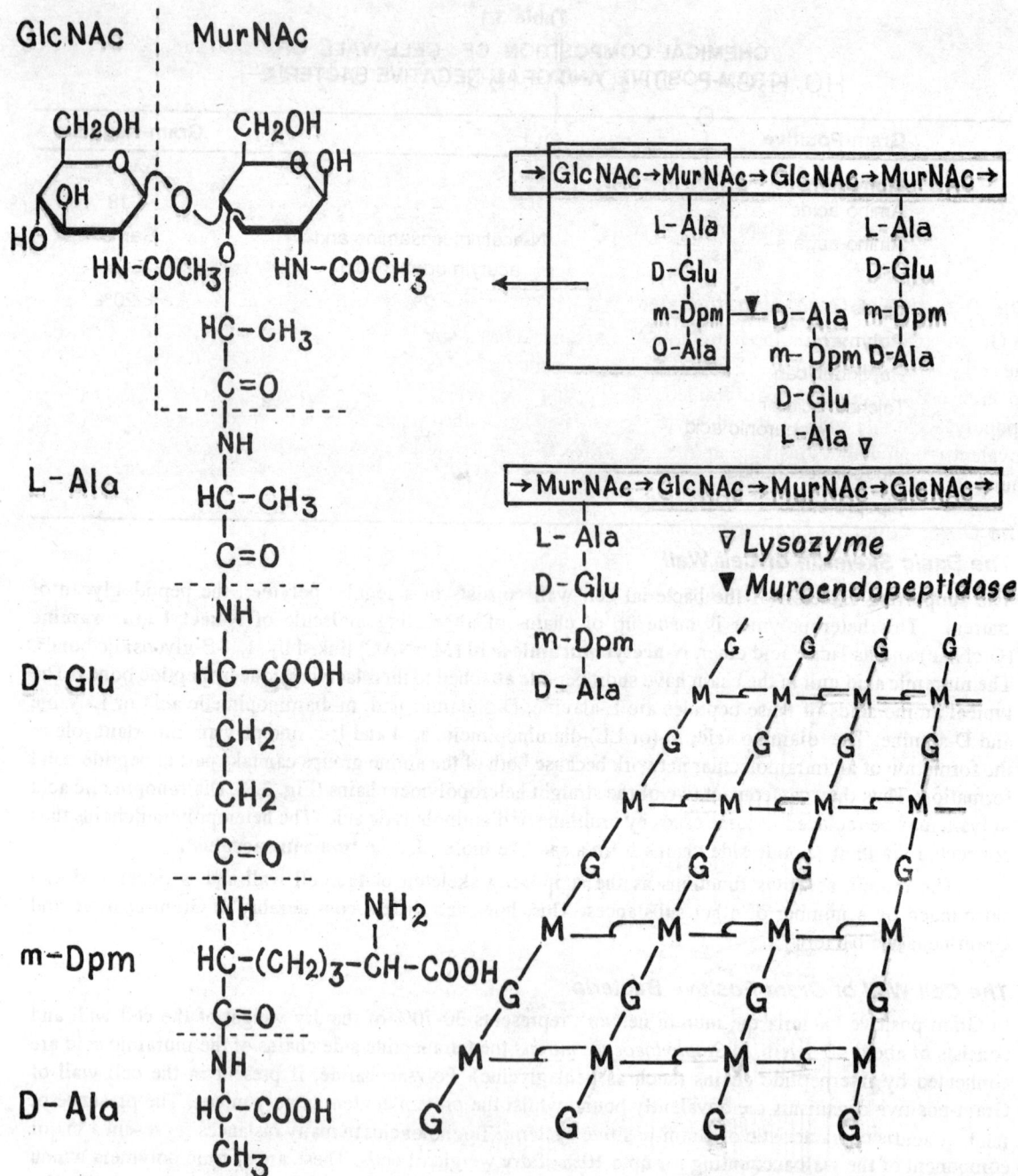

Fig. 3.8 Murein structure in *E. coli.*

Alternating chains of N-acetyl glucosamine and N-acetyl muramic acid are connected by peptide bonds that can be split by lysozyme and muroendopeptidase.

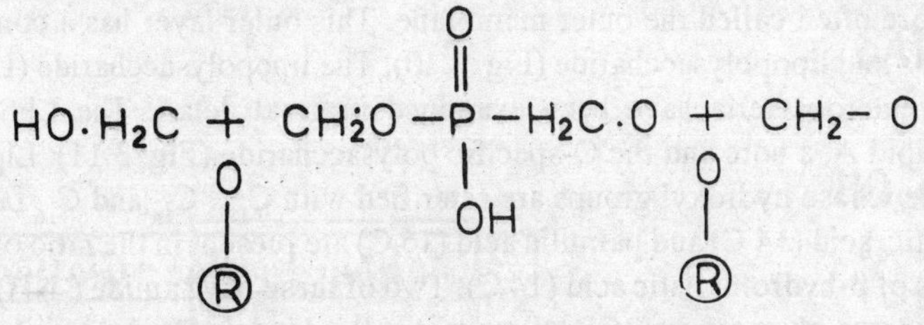

$$\text{HO·H}_2\text{C} + \text{CH}_2\text{O}-\overset{\displaystyle O}{\underset{\displaystyle OH}{\overset{\|}{P}}}-\text{H}_2\text{C·O} + \text{CH}_2-\text{O}-$$

Fig. 3.9   Structure of 1, 3-polyglycerol phosphate teichoic acid. R = H or D-alanine or sugar residue.

The Cell Wall of Gram-Negative Bacteria

In Gram-negative bacteria the murein network is present as a single layer and represents less than 10% of the cell wall dry weight (e.g., *E. coli*). The murein contains no lysine, only m-diaminopimelic acid and has no inter-peptide bridge. Apart from the supporting skeleton there are large quantities of lipoprotein, lipopolysaccharide and other lipids which appear attached to the outer surface of murein skeleton. They are covalently bound and constitute upto 80% of the cell wall dry weight. So far, no techoic acids have been found in Gram-negative organism.

The Outer Layer of Gram-Negative Cell Wall

In Gram-negative bacteria the monolayer or bilayer of the murein sacculus is surrounded by an outer cell wall layer or envelope. In thin section of electron micrograph this appears similar to cytoplasmic

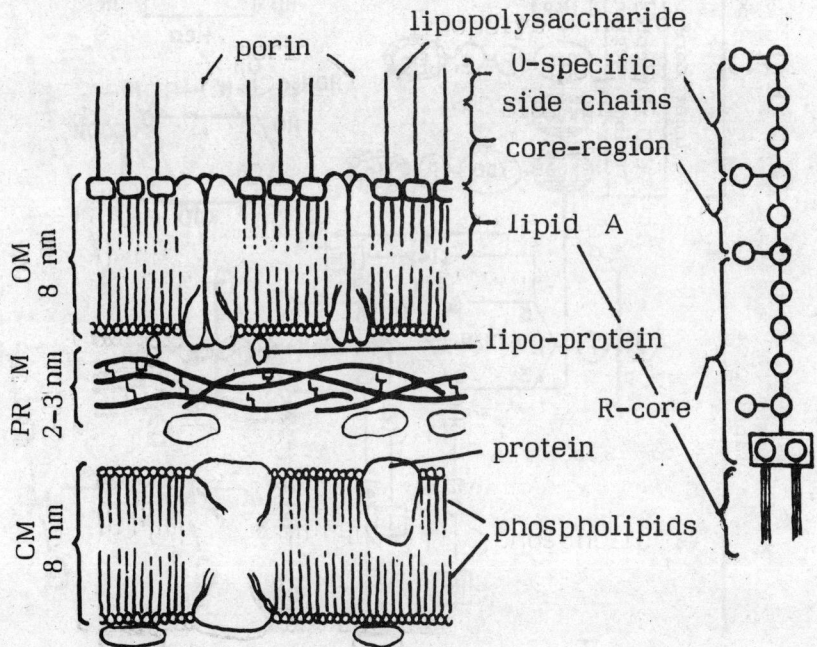

Fig. 3.10 Outer membrane of Gram-negative cell.

The murein layer (M) surrounds the cytoplasmic membrane (CM) and periplasmic space (PR). The lipophilic ends of the lipo-proteins are embedded in a lipid bilayer which contains phospholipids and the lipid A zone of lipopolysaccharides. (Right) detailed structure of lipopolysaccharide. Glc, glucose; Glc-N, glucosamine; Gal, galactose; Hep, heptose; KDO, 2-Keto-3-deoxyoctanoic acid.

membrane and is therefore often called the outer membrane. This outer layer has a complex composition of proteins, phospholipids and lipopolysaccharide (Fig. 3.10). The lipopolysaccharide (LPS) of *Salmonella typhimurium* and other enterobacteria have been examined in great detail. The LPS consists of three complex constituents: Lipid A, a core and the O-specific polysaccharide (Fig. 3.11). Lipid A consists of a glucosamine dissacharide whose hydroxyl groups are esterified with C_{12}, C_{14} and C_{16} fatty acids. Of these lauric acid (12 C), myristic acid (14 C) and palmitic acid (16 C) are present in the ratio of 1:1:1. In addition there are three molecules of β-hydromyristic acid (14 C). Two of these have amide (-NH) linkage with each of the amino groups of the two glucosamine (GLcN) units, the third is esterified through its hydroxyl group to myristic acid.

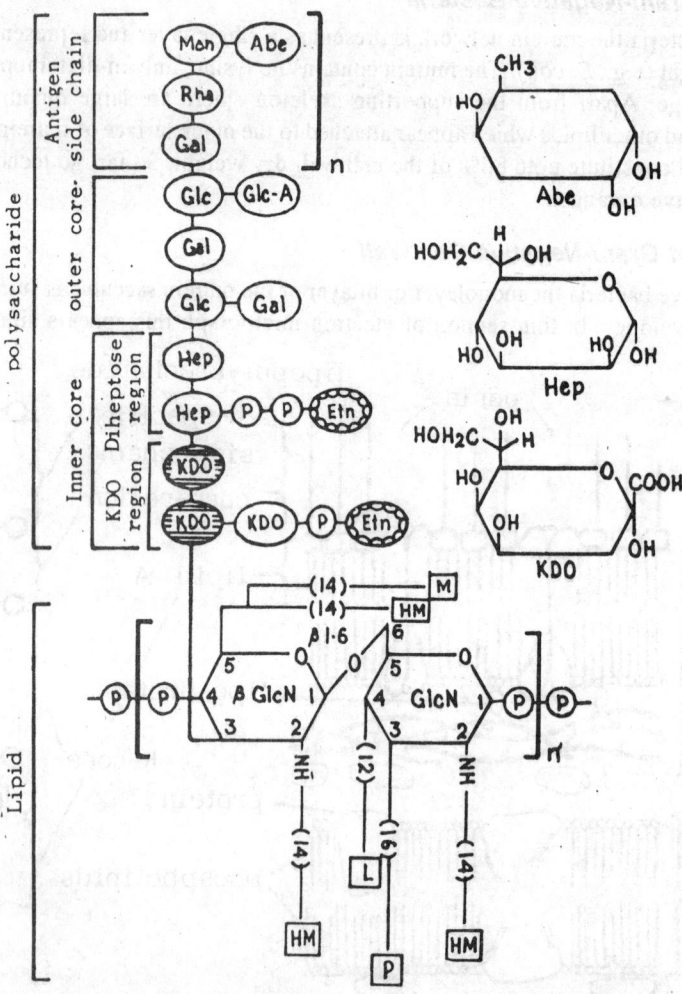

Fig. 3.11 Diagramatic representation of a unit of lipopolysaccharide molecule. Abe - abequose; Etn - ethanolamine; Gal - galactase; GlcA - glucose acetyl; GlcN - glucosamine; Hep -L-glycero-D-mannoheptose; KDO - 2-Keto-3-deoxyoctonate, HM - B-hydroxymyristic acid (14C), L - lauric acid (12C); M - myristic acid (14C), P - palmitic acid; P - phosphate.

The lipopolysaccharides have achieved great significance in diagnostic bacteriology. Different strains of *Salmonella typhimurium, Shigella dysenteriae* and other causative organisms of intestinal infection are differentiated by their O-specific side chains of lipopolysaccharide. Many bacterial strains grow on agar medium as smooth, shiny colonies (S-forms). Their O-specific polysaccharides in the cell surface apparently retain water. These S-forms may mutate spontaneously to R-forms which show flat, rough colonies. In the animal host these bacteria are very resistant to phagocytosis and hence virulent. The lipopolysaccharide has most effective endotoxin properties causing fever and diarrhoea to the host.

The Function of the Outer membrane

The outer membrane of Gram-negative bacteria has important physiological as well as mechanical function. Along with the murein layer the outer membrane proteins contribute significantly to the structural integrity of the cell. The lipid bilayer consisting of Lipid A and phospholipids having transmembrane proteins are thought to constitute the water filled channel or hydrophilic pores in the lipophilic membrane and are called **porins**. They act as diffusion barrier which allow low molecular weight substance to enter the cell. In between murein layer and the cytoplasmic membrane is the periplasmic space. It contains enzymic protein such as depolymerases (proteinases and nuclease; the restriction enzyme, ECO R_1), peripheral protein of the cytoplasmic membrane and the so called binding proteins. These binding proteins function in the transport of some substrate into the cytoplasm and are also receptors for chemotactic stimuli for various nutrients, bacteriophages and colicines. The periplasmic space also plays an important role in osmoregulation.

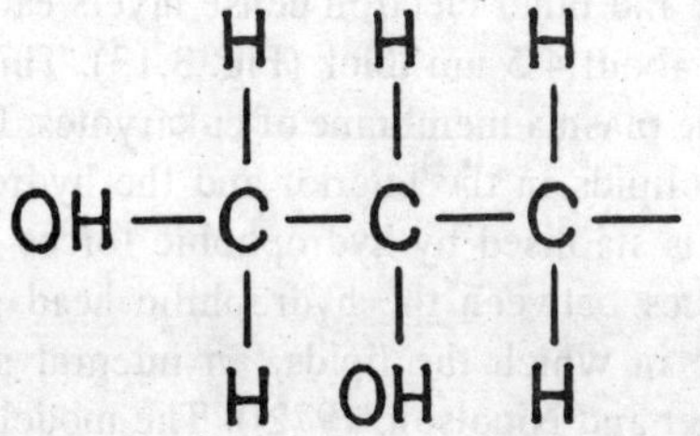

PHOSPHATIDYL GLYCEROL 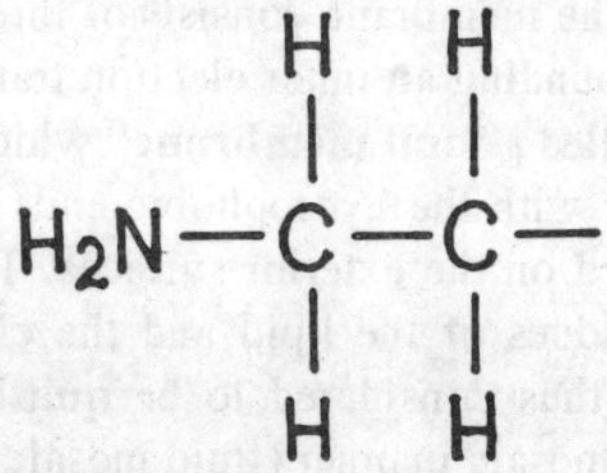   PHOSPHATIDYL ETHANOLAMINE

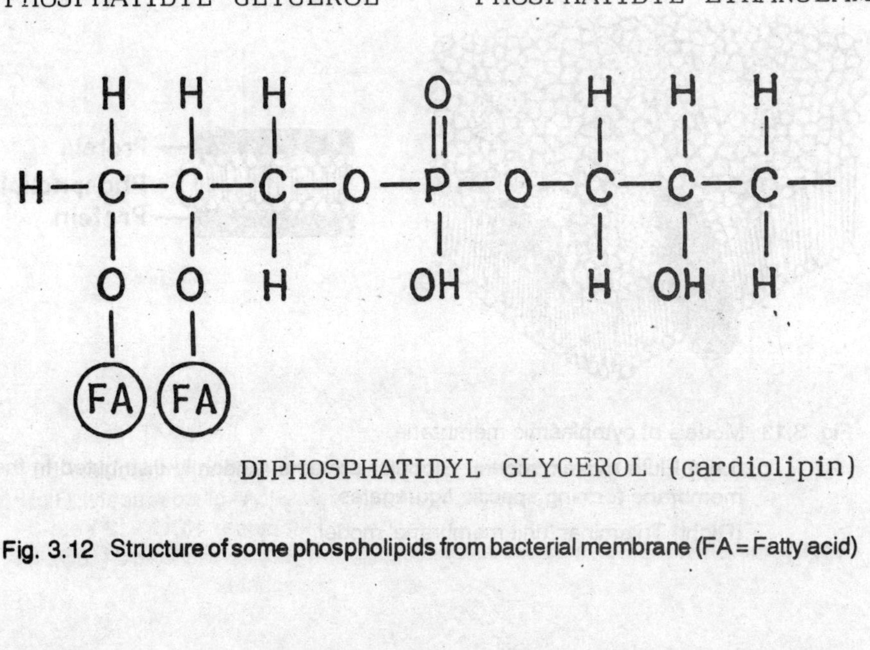

DIPHOSPHATIDYL GLYCEROL (cardiolipin)

Fig. 3.12 Structure of some phospholipids from bacterial membrane (FA = Fatty acid)

CYTOPLASMIC MEMBRANE

Immediately beneath the cell wall is a thin membrane or covering called the cytoplasmic membrane also referred to as **periplasmic membrane** or the plasma membrane. It is about 7.5 nm thick, accounts for about 8-15% of the cell's dry weight and consists mostly of lipid (16-29%) and protein (40-75%) with little amount of nucleic acid and carbohydrate. The predominant lipids in the cytoplasmic membrane are phospholipids, in particular, phosphatidylglycerol and/or phosphatidyl ethamolamine (Fig. 3.12). Membranes of Gram-positive bacteria characteristically contain one or both of these together with two or three more closely related compounds such as phosphatidic acid and diphosphatidylglycerol whereas that of Gram-negative bacteria typically contain only one or two lipids. Even the lipid composition and fatty acid moieties vary considerably in Gram-positive and Gram-negative bacteria. In the former, usually branched chain saturated types of fatty acids are found whereas in Gram-negative species mixture of saturated, unsaturated and straight chain varieties are seen.

Besides lipids, two broad categories of protein are recognised in the cytoplasmic membrane in contrast to the outer membrane of Gram-negative cell wall—peripheral protein and integral protein. The peripheral protein are soluble and readily dissociate from the membrane. The integral proteins are relatively insoluble and dissociate with difficulty. The peripheral proteins are entirely outside the lipid bilayer whereas the integral protein is considered as floating in the membrane matrix, some completely traversing it, whilst others are partially immersed.

In electron micrographs of ultrathin sections from osmium tetroxide treated cells the bacterial cytoplasmic membrane appears to consist two dense lines between the cytoplasm and the cell wall. This suggests that the membrane consists of three layers, outer and inner electron dense layers each about 2-3 nm thick surrounding an inner electron transparent space about 4-5 nm thick (Fig. 3.13). This composite structure is called a **'unit membrane'** which resembles the plasma membrane of eukaryotes. It consists of a lipid bilayer with the hydrophobic ends of the phospholipids in the interior and the hydrophilic head groups exposed on the exterior surfaces. The membrane is stablised by hydrophobic forces between the fatty acid residues of the lipid and the electrostatic forces between the hydrophilic head groups. The membrane is thus considered to be **quasifluid** structure in which the lipids, an integral proteins, are arranged in a mosaic manner (fluid mosaic model of Singer and Nicolson, 1972)). The model is currently widely accepted as a basis for explanation of the permeability barrier functions of the cytoplasmic

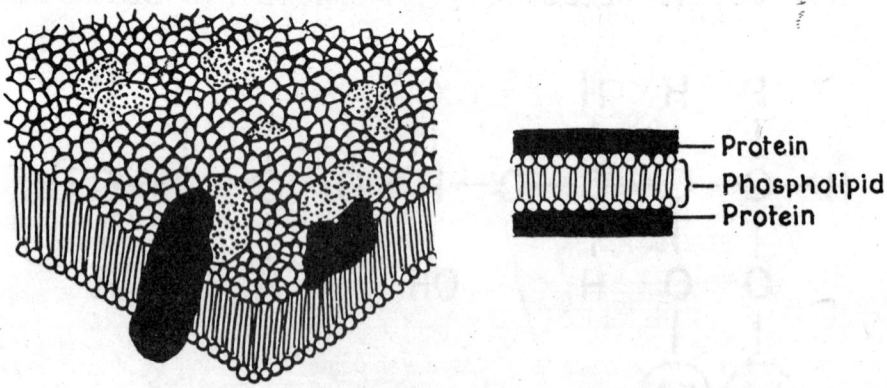

Fig. 3.13 Models of cytoplasmic membrane.

(Left) Fluid mosaic model. Globular proteins randomly distributed in the membrane forming specific aggregates

(Right) Trilaminar 'unit membrane' model.

membrane. However, it is incompatible with some of its other properties.

The cytoplasmic membrane has many important metabolic functions. (i) It constitutes the outer barrier of the cell and exerts control over the influx and efflux of materials (ii) It is the site of active transport mechanisms and of substrate specific permease system (iii) The continuous lipid bilayers traversd by bridging proteins form pore through which regulated transport of substances take place. (iv) The enzymes of electron transport and oxidative phosphorylation which in eukaryotes are localised in the mitochondrial membranes, are part of bacterial cytoplasmic membrane. Thus cytochrome, iron-sulphur proteins and other components of electron transport chain are found solely in the cytoplasmic membrane. (v) Various biosynthetic processes like synthesis of cell wall and capsule components and the secretion of exoenzymes are most probably the membrane functions (vi) The replication centre of DNA is also thought to be localised on the membrane (vii) The membrane also provides the anchorage site of flagella; it also contains the attachment sites for chromosomal and plasmid DNA (viii) The photosynthetic apparatus of purple bacteria and the components for the control of chemotoxins appear to be located in the membrane.

INTRA-CYTOPLASMIC MEMBRANE AND LAMELLAE

A distinguishing feature of the prokaryotes is their lack of intracellular membrane structure. In many Gram-positive bacteria, however, invagination of the cytoplasmic membrane occurs (Fig. 3.14). These organelles are usually referred to as **mesosome** to which many physiological functions are attributed. These mesosomes consist of vesicles tubules or lamellar whorls filling the invagination. The vesicular type and whorl type appear to be the most commonly seen forms. However, there is considerable doubt about their true morphological nature; most probably they are the artefacts of preparation. Still many workers believe that these are involved in functions of reproductive and metabolic processes. For example, they are frequently associated with septum formation during the process of bacterial cell division. These also take part in the replication of bacterial nuclear material. Certain enzymatic processes e.g., electron transport are likewise associated with mesosome material.

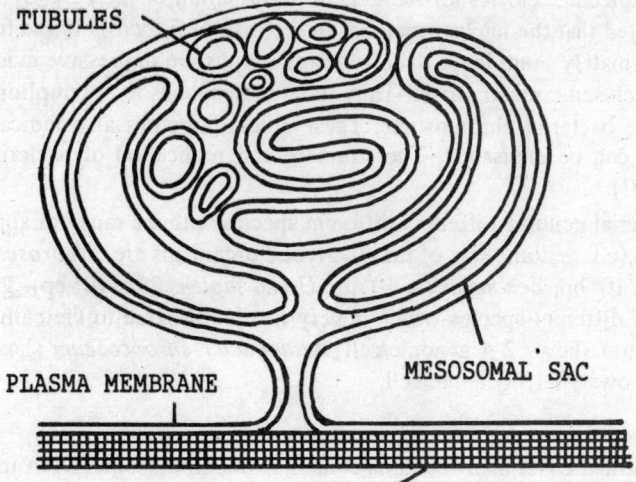

Fig. 3.14 A bacterial mesosome as demonstrated in an electron micrograph of an ultrathin section.

Phototrophic purple bacteria are especially rich in intra-cytoplasmic membrane systems. In ultrathin sections these can be seen as tubules, vesicles and plates. In *Photospirillum rubrum* and *Chromatium* sp. the cell lumen appears almost completely filled with closely packed spherical vesicles. These vesicles seem to be formed by invagination and tubular accretion of the cytoplasmic membrane. Upon cell breakage and as a result of homogenising procedures, these structures are liberated as isolated vesicles and have been called **chromatophores**. In other purple bacteria the vesicular bodies are flattened and stacked into regular plates. They are often called **thylakoid** plates; analogy to the structure of chloroplasts in green plants.

THE BACTERIAL NUCLEUS

Bacterial cells do not contain the nucleus characteristic of the cells of higher plants and animals. They do, however, contain bodies within the cytoplasm that are regarded as a nuclear structure. Since it is not a discrete nucleus, it is often designated as **chromatin body, nucleoid, nuclear equivalent** or **bacterial chromosome**. For our convenience, we use the term nucleus bearing in mind that it is a primitive form of nucleus. The chromatin bodies or nuclei are more or less centrally located in resting cells and are spherical or oval or rod-shaped. They lack a bounding membrane. However, it is a definite structure, readily distinguishable from the remainder of the cell contents by the use of nuclear stains. These are of two general types : (i) solid structures forming bars or H, V, or butterfly shape, shown particularly by staining and (ii) small granules enmeshed in fine strands best seen in photomicrograph. The nuclei vary in dimensions between species and within the same species at different stages. Resting cells of some species of *Staphylococcus* possess chromatin bodies about 0.4 μ in diameter whereas in growing cells they enlarge to about 0.5 to 0.8 μ. The small size of bacteria and the presence of two types of nucleic acid has made the cytochemical study of bacterial nuclear material very difficult. However, classical cytological methods and the developments of ultrathin section techniques allied to electron microscopy eventually established that bacteria contain DNA that is not diffusely distributed within the cytoplasm but occupies discrete region and divides before each cell division.

Bacterial nucleus appear to be composed of fine fibrils of DNA or deoxyribonucleo-protein, 0.3 to 0.4 μm in diameter. In Gram-negative bacteria like *E. coli* and *Salmonella typhimurium* these fibrils are arranged in delicate but compact whorl whereas in Gram-positive bacteria (e.g., cocci and bacilli) the dense fibres are aligned in an almost parallel pattern. The DNA constitutes a single chromosome and the single two stranded molecule carries all the genetic information of the cell. By using autoradiography Cairns first demonstrated that the nuclear material in the case of *E. coli* is in the form of a single closed, circular strand approximately 1 nm long. Such autoradiographs are impressive evidence for the existence of bacterial DNA as a closed circular strand. This strand corresponds to a **'coupling group'** in the genetic sense and is called the bacterial chromosome. These autoradiographs also indicate how the process of chromosome division can be envisaged. The structure and replication of bacterial DNA are described elsewhere (Chapter VII).

The size of bacterial genome differs in different species within a range of approximately 0.8 x 10⁶ 8 x 10⁶ base pairs (bp) (c.f. genome size of the eukaryotic organisms are : *Neurospora crassa* 19 x 10⁶ bp; *Aspergillus niger* 40 x 10⁶ bp; *Zea mays* 7 x 10⁹ bp; *Homo sapiens* 29 x 10⁹ bp). The number of genomes per cell also differs in different species which is very much attributed to the cultural conditions. *E. coli* growing in batch culture shows 2-4 genome/cell; *Azotobacter chroococcum* shows 20-25 genome/cell; *Desulfovibrio gigas* shows 10-15 genome/cell.

PLASMID

In addition to chromosomal DNA many bacteria contain extrachromosomal DNA in closed circular double stranded form. This unit of genetic material capable of independent replication is called **plasmid**. Plasmids are found in almost every known type of bacterial cell. Most of the plamids are dispensible i.e., they are not required for the survival of cell in which they reside. In many cases, however, they are essential under certain environmental conditions such as in the presence of an antibiotic. Their roles in modern biotechnology have opened a new era of genetic engineering. Detailed account of plasmid has been presented separately in chapter VII.

CYTOPLASMIC INCLUSIONS

The rapidly growing bacterial cell contains numerous cytoplasmic inclusions. Although there is considerable variations in cytoplasmic inclusions in different species, the approximate composition of a representative cell is indicated in Table 3.2.

Table -3.2
APPROXIMATE COMPOSITION OF A BACTERIAL CELL

Water	70%
Dry weight	30%
DNA	3% (MW = 2 x 10^9)
RNA	12%
PROTEIN	70% found in
Ribosome (10,000)	(RNA-protein particle M.W. 3x10^6)
Enzyme	
Surface structure	
POLYSACCHARIDES	5%
LIPID	6%
PHOSPHOLIPID	4%

The ribosomes are the sites of protein synthesis. In electron micrographs they are seen as particles in the cytoplasm. Bacterial ribosomes are about 16 x 18 nm in size and contain about 80-85% of the bacterial RNA. Since intact bacterial ribosome sediment in ultracentrifuge with a sedimentation velocity of about 70 *svedberg unit*, they are referred to as **70 S ribosomes**. Ribosomes consist of two subunits. In bacteria these are 30 S and 50 S particles which combine to give the 70 S ribosome (Fig. 3.15). Bacterial cell contains about 5000-50,000 ribosomes which resemble those of mitochondria and chloroplast in size and other properties.

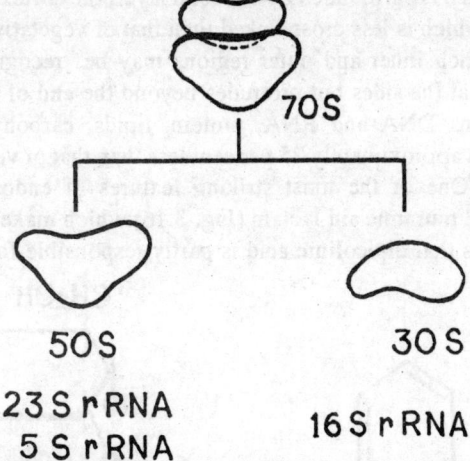

Fig. 3.15 Composition of bacterial ribosome.

All the three kinds of RNA - messenger RNA, ribosomal RNA and transfer RNA are met in bacterial cell. Messenger RNA (*m* RNA) has a molecular weight of 1,00,000; ribosomal RNA (RNA) associated with protein is known as true ribosome and transfer RNA (*t*RNA) (also called soluble RNA) are relatively small being composed of 80 to 100 nucleotide units and having a molecular weight between 25,000-30,000.

Many of other inclusions in bacterial cells act as reserves of energy and/or elements and their formation depends largely on growth conditions. Usually these bodies consist of polymeic, osmotically inert materials. Some of these bodies are laid down in the cytoplasm without any bounding membrane. Polyphosphate granules (Volutin : metachromatic) are one example which lack a surrounding membrane in *Corynebacterium diphtheriae*. Similarly polyglucoside granules in most instances are non-membranous but some deposits of this material are bounded by a membrane which may consist entirely of protein. As per their reaction with iodine solutions, polyglucoside granules are often referred to as glycogen (stain red-brown) and starch (stain blue). Other membrane enclosed deposits are poly-β-hydroxybutyrate (PBH) and sulphur. PBH granules are widely distributed among bacteria where intracellular sulphur granules are restricted in the purple sulphur bacteria or colourless sulphur bacteria (e.g., *Beggiatoa*). Some aquatic prokaryotes contain gas vacuoles whose component vesicles are bounded by a membrane consisting of protein. In some Gram-positive bacteria lipid globules are also present. They can be stained with lipophilic dyes like Sudan-III or Sudan black-B and are also visible in unstained preparation due to their high refractivity.

ENDOSPORES AND OTHER PERSISTENT FORMS

A small group of bacteria are capable of producing endospores. Endospores are the most resistant of all living bodies to heat, desiccation and toxic chemicals. Whilst most of bacterial cells are killed by heating at 80°C for 10 minutes (Pasteurisation) the heat resistant endospores can withstand for greater thermal exposure; some are resistant to boiling water.

With one exception, all spore formers are rod-shaped, Gram-positive bacteria. The members of genus *Bacillus* are all strict or facultative aerobes. The anaerobic spore formers are the genera of *Clostridium* and *Desulfotomaculum*. Besides, *Sporolactobacillus, Sporosarcina, Azotobacter* (cyst formation), *Myxobacteria* (microcyst) are also known to produce endospores or alike structures. These spores are easily recognised because of their high refractivity. In doubtful cases, specific spore stains (e.g., malachite green or carbolfuchsin) are used to determine the presence of true endospores.

Endospore typically have their nuclear material arranged as filament. Around their cytoplasm is a thin membrane, spore wall, which is surrounded by a second layer, the cortex. This is a substantial structure containing the peptidoglycan which is less cross linked than that of vegetative cell. Outside the cortex lies the stratified spore coat in which inner and outer regions may be recognised. The outermost layer is **exosporium** which fits snugly at the sides but protrudes beyond the end of the spores. Chemical analysis reveals that endospore contains DNA and RNA, protein, lipids, carbohydrate, various enzymes and minerals. Their water content is approximately 25 per cent less than that of vegetative cells. In addition, the pores contain more calcium. One of the most striking features of endospore is the presence of the compound, dipicolinic acid and muramic aid lactam (Fig. 3.16) which make up 5 to 15 per cent of the dry weight of the spores. It appears that dipicolinic acid is partly responsible for spore resistance.

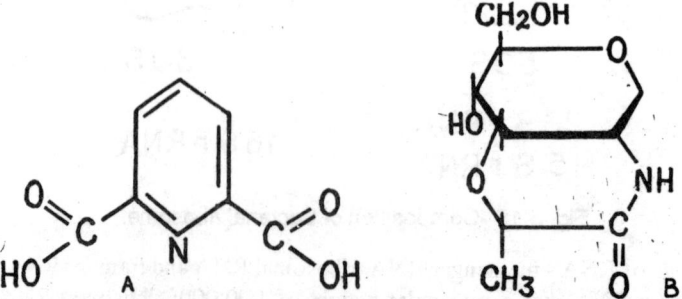

Fig. 3.16 Chemical structures of dipicolinic acid (A) and muramic acid lactam (B) of bacterial endospore.

Endospores arise intracellularly starting with the accumulation of protein rich material. Numerous metabolic conversions occur at the expense of storage materials. In fact spores are not an obligate part of the life-cycle in bacteria; under favourable nutritional conditions even spore forming bacilli can continue to grow indefinitely by division of vegetative cells. Spore formation is initiated when nutrients are exhausted or unfavourable metabolic products accumulate. Spores are liberated upon autolysis of the vegetative cell. The mature spores have no demonstrable metabolic activity and exhibit a high degree of resistance to heat, radiation and chemicals. Spores enable bacteria to survive in a latent state for a long time.

Besides endospores a few bacteria produce other kinds of peristent forms. The formation of exospores has been observed in the methane utilising bacterium *Methylosinus trichosporium* by budding of vegetative cell. Some bacteria form spherical, thick walled cells that are called cysts. This formation of **cysts** occurs when nutrients are exhausted. The cysts of *Azotobacter* and *Methylocystis* species are resistant to drying, mechanical stress, and radiation but not to heat. A similar transformation occurs in the production of myxospores from the rod shaped vegetative cells of *Myxococcus* and *Sporocytophaga*.

PIGMENTS

Many bacterial colonies show distinct colouration because their cells are pigmented. Coloured pigments are easy to recognise and identify. Many of the pigments are derivatives of certain groups of chemicals such as carotenoids, phenazine or pyrollic dyes, azaquinone, anthocynates etc. Some of the pigments produced by bacteria are presented in Fig. 3.17.

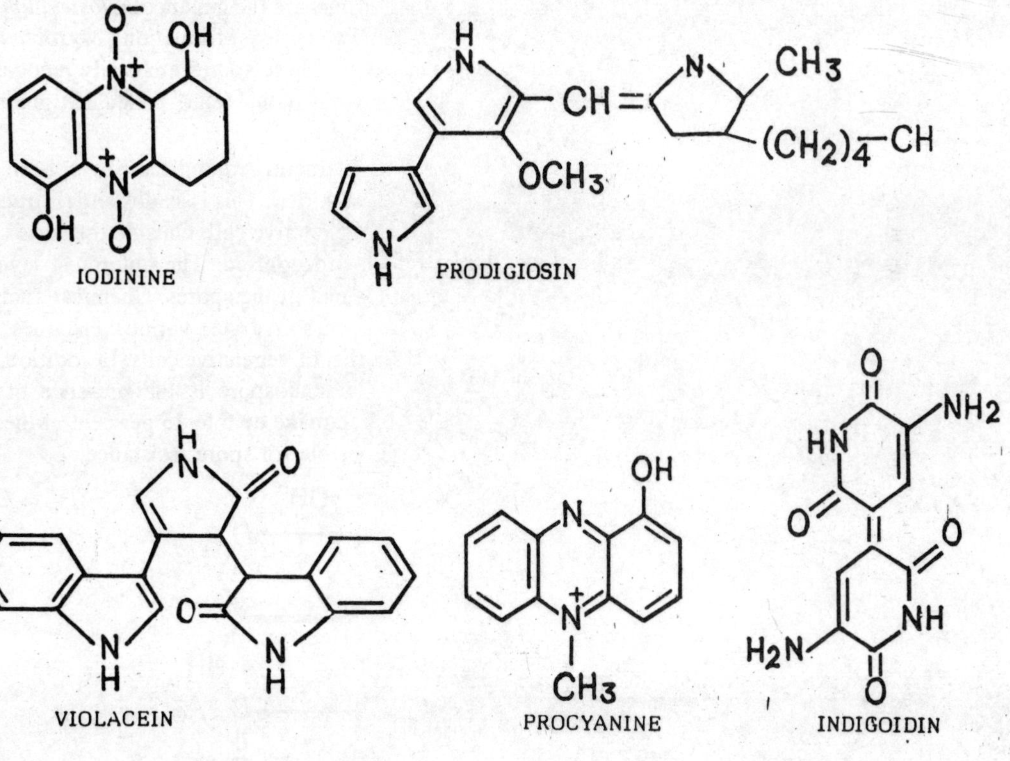

Fig. 3.17 Some of the pigments produced by bacteria.

SUGGESTED READINGS

Buchanan, R.E. and Gibbons, N.E. (1974). Bergey's Manual of Determinative Bacteriology. 8th Ed. Williams and Wilkins, Baltimore.

Hawker, L.E. and A.H. Linton (1979). Micro-organisms : Function, Form and Environment. Edward Arnold Publishers Ltd. London.

Kandler, O. (1982). Cell wall structures and their phylogenetic implication. Symp. Soc. General Microbiol. **32**: 33.

Ottow, J.C.G (1975). Ecology, physiology and genetics of fimbria and pili. Rev. Microbiol, **29**: 79-108.

Pollack, M.R. and M.H. Richmond (eds.) (1965). Function and structure in Microorganisms. Cambridge, New York.

Schlegel, H.G. (1986). General Microbiology. Cambridge University Press, Cambridge.

Schleifer, K.H. and Kansler, O. (1972). Peptidoglycan types of bacterial cell-walls and their taxonomic implications. Bact. Rev. **36**: 407-477.

Stanier, R.Y., Rogers, H.J. and Ward, J.B. (1978) Relation between structure and function in the Prokaryotic cell. 28th Symp. Soc. Gen. Microbiol. Cambridge University Press, Cambridge.

CLASSIFICATION AND DIVERSITY OF PROKARYOTES

4

The prokaryotes include a variety of morphologically and physiologically diverse forms that manifest an exta-ordinary variation in their metabolic mechanisms ranging from autotrophic, photosynthetic to heterotrophic modes of nutrition. Classification of these prokaryotes is somewhat different from that of eukaryotes where grouping is based on the ability of two organisms to sexually produce fertile offspring with each other and not with members of other species. The classification of prokaryotes is based on a number of arbitrary criteria which are believed to indicate whether organisms are genetically related to one another or not. Such criteria include Gram-stain reaction, morphology, the ability to produce endospores and mechanisms of metabolism, motility and mode of reproduction. These criteria are not necessarily an accurate reflection of genetic relatedness. However, because most of these morphological and biochemical properties are relatively easy to determine in the laboratory, these have proven to be of practical significance for segreggating the related forms.

GENETIC RELATEDNESS

In recent decades concept of DNA similarity has been taken up as an important criterion which has influenced the classical taxonomy. It is the ultimate indicator of genetic relatedness. Prokaryotic cells are easiest to analyse for the DNA similarities because they have single chromosome.

Genetic relatedness of organisms is often measured by comparing the size of chromosome, chemical composition of the DNA (G+C content) and DNA homology. The degree of DNA homology is the best indicator of genetic similarities between two organisms. As per rules laid down by the International Judicial Commission, organisms that are about 40 to 60 per cent related for DNA homology are considered members of a single genus. Those organisms that are more than 70 per cent related are assigned to a single forms. Similarities below 30 per cent indicate the unrelated forms. If an organism is too different to be assigned to any existing genus or species it is declared a new organism and given its own name.

BERGEY'S MANUAL

The most standard reference of descriptive bacteriology is 'Bergey's Manual of Determinative Bacteriology, 8th edition (1974)' which contains names, description of morphological and physiological properties with citation of relevant literature and determinative keys for classification of new isolate. The manual has been a practical guide for the identification of bacteria ever since its first publication in 1923. The classification scheme is based on the available genetic data that is related to the morphological and biochemical tests. The key describes the morphological characters i.e., their shape - rod-shaped, coccoid or spirillar; presence/absence of capsules; occurrence i.e., single or in groups; presence/absence of flagella and their orientation; spore producing capacity and response to Gram-stain. These descriptions are complemented by enumeration of physiological and biochemical characters i.e., (i) whether the cells can grow under aerobic, anaerobic or both the conditions (ii) whether energy is derived by respiration, fermentation or photosynthesis (iii) utilisable nutrients (iv) temperature and pH dependence (v) habitat (vi) symbiotic or parasitic relationship with other organisms (vii) cell inclusions, pigmentation and capsular materials (viii) composition of cell wall component (ix) serological properties (surface antigens, homologous proteins) (x) base composition of DNA (CG content) (xi) DNA-DNA hybridisation, transformability by interspecies transfer (xii) sequence of 16 S or 5 S rRNA (xiii) antibiotic sensitivities.

Bergey's Manual of Determinative Becteriology (1974) classifies bacteria into 19 groups. Formerly it was believed that bacterial world could be classified to reveal natural or evolutionary relationships. However, microbiologists particularly bacteriologists later on subscribed that the hierarchical scheme was purely theoretical because adequate information was not available to develop a natural classification scheme that reflects evolutionary relationships. The Manual (8th ed.) was therefore, presented based on a few readily determinative criteria. The thumbnail sketch of these 19 parts is presented in Table 4.1.

In the year 1983 the ninth edition of Bergey's Manual was released. In this edition the prokaryotes have been separated into four divisions, based on the nature of the cell wall. These are **Graculicutes** (*graculo*, thin; *cutes*, skin), Gram-negative bacteria; **Firmicutes** (*firmi*, strong), Gram-positive bacteria; **Tenericutes** (*tener*, soft); wall-less bacteria; and **Mendosicutes** (*Mendos*, faulty), bacteria possessing cell wall that lacks peptidoglycan. However, for all practical purposes, the eighth edition still continues to be the major guideline for classifying and identifying bacteria. It is widely accepted by the bacteriologists.

Despite presenting an illustrative account of the diversity of the prokaryotes, Bergey's Manual (1974) does not accommodate cyanobacteria, nor the archaebacteria, possibly for the reason that their establishment as separate entities were not known at the time of its publication. Many recent discoveries make it highly probable that a number of modifications and additions will need to be introduced in the manual. In the following section a novel approach to describe the prokaryotes has been used. Prokaryotes have been arranged according to their (i) shape (cocci, rods, spirilla), (ii) Gram-stain reaction and (iii) relationship to oxygen (aerobic or anaerobic) (see table 4.2). The group numbers of Bergey's Manual (8th edition) have been incorporated for the convenience of the students. Cyanobacteria has been treated as a separate Group 20. Those bacteria that can not be assigned to any of the above three basic catergories are listed here as **'Large Special Group'**. A brief resume of each group is given as below.

COCCI

1. GRAM-POSITIVE COCCI (GROUP-14)

This group of bacteria inculdes three families; Micrococcaceae, Streptococcaceae and Peptococcaceae. Members of former two families are aerobic or facultatively anaerobic Gram-positive cocci. *Micrococcus* is common in soils and fresh water and exhibits oxidative metabolism (G+C ratio = 66-75 mole %). *Staphylococcus* is facultative anaerobes, pathogenic on warm blood animals (G+C ratio = 30-40 mole %). *S. aureus* produces toxin and exoenzymes and is a pus former. In general Gram-positive cocci are referred to as lactic acid bacteria that include *Streptococcus, Leuconostoc* and *Pediococcus*. Besides, some obligate anaerobic forms e.g., *Sarcina, Peptococcus, Ruminococcus* are also included in this group.

Table 4.1

THE MAJOR GROUPS OF BACTERIA (ILLUSTRATED IN BERGEY'S MANAUAL)

	Group		Important Representative Members
Part-1	Phototrophic bacteria		*Rhodospirillum, Rhodopseudomonas, Chromatium, Chlorobium*
Part-2	The gliding bacteria		*Myxococcus, Beggiatoa, Leucothrix, Polyangium, Cytophaga*
Part-3	The sheathed bacteria	A.	*Sphaerotilus, Leptothrix, Streptothrix, Crenothrix*
Part-4	Budding and/or append aged bacteria		*Caulobacter, Hyphomonas, Hyphomicrobium, Plandomyces*
Part-5	Spirochaetes	A & An.	*Spirochaeta, Cristispira, Treponema, Borrelia*
Part-6	Spiral and curved bacteria	A.	*Spirillum, Bdellovibrio, Azospirillum, Campylobacter*
Part-7	Gram-negative aerobic rods	A.	*Pseudomonas, Xanthomonas, Acetobacter, Azotobacter, Rhizobium, Beijerinckia*
Part-8	Gram-negative aerobic facultatively anaerobic rods		*Escherichia, Klebsiella, Salmonella, Shigella, Yersinia, Vibrio*
Part-9	Gram-negative anaerobic bacteria	An.	*Bacteroid, Fusobacterium, Leptotrichia*
Part-10	Gram-negative cocci and coccobacilli	A.	*Neissaria, Moraxella, Paracoccus, Lampropedia*
Part-11	Gram-negative anaerobic cocci	An.	*Veillonella, Megasphaera , Acidaminococcus*
Part-12	Gram-negative chemo-lithotrophic bacteria	A.	*Nitrobacter, Nitrosomonas, Nitrococcus, Thiobacillus*
Part-13	Methane producing bacteria	A.	*Halobacterium, Halococcus*
		An.	*Methanobacterium, Methanosarcina, Methanothrix*
Part-14	Gram-positive cocci	A.	*Micrococcus, Staphylococcus, Streptococcus, Luconostoc*
		An.	*Ruminococcus, Sarcina*
Part-15	Endospore forming rods and cocci	A.	*Bacillus, Sporosarcina*
		An.	*Clostridium, Oscillospira*
Part-16	Gram-positive asporogenous rods-shaped bacteria	A.	*Lactobacillus, Listeria, Erysipelothrix*
Part-17	Actinomycetes and related organisms	A.	*Corynebacterium, Arthrobacter, Brevibacterium, Mycobacterium, Nocardia, Actinomyces*
Part-18	Rickettsias		*Rickettsia, Coxiella, Chlamydia*
Part-19	Mycoplasma		*Mycoplasma, Acholeplasma, Spiroplasma.*

(A = Aerobes ; An = Anaerobes)

2. GRAM-NEGATIVE COCCI (GROUP-10 & 11)

This group contains some cocci and very short rods that are non-motile. Included among are aerobe and anaerobe pathogens, soil bacteria and genera that inhabit the intestinal tract and mucous membrane of various animals.

Table 4.2

GENERAL SCHEME FOR THE CLASSIFICATION OF PROKARYOTES

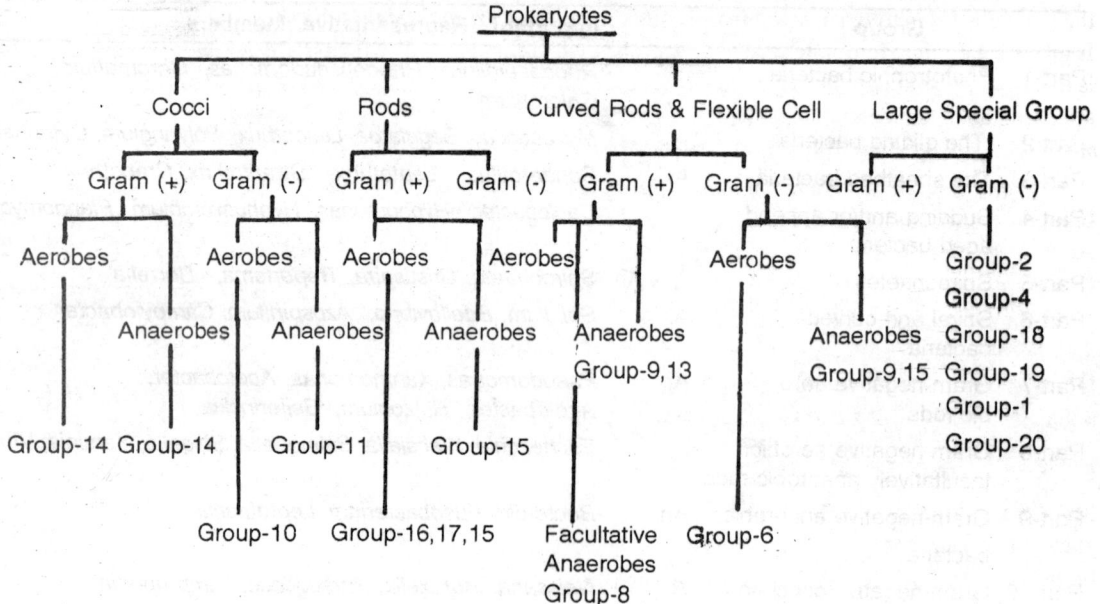

Aerobic cocci (Group-10) - The representative genera are *Neisseria, Branhamella, Moraxella* and *Acinetobacter*; G+C ratio is 39-52 moles %. The genus *Neisseria* is oxidase-positive and contains several animal and human pathogens. The species *N. gonorrhoeae* causes gonorrhoea in human beings. *N. meningitidis* inhabits the nasopharyngeal tract and produces meningitis leading to inflammation. *Moraxella* is also oxidase-positive and very sensitive to penicillin. *Acinetobacter* is an oxidase negative bacterium.

Anaerobic cocci (Group-11) - The best known Gram-negative anaerobic cocci are *Veillonella alcalescens* and *Megasphaera elsdenii*. These are parasitic bacteria found in the alimentary tract of warm blooded animals (G+C ratio = 40-47 moles %).

RODS

3. GRAM POSITIVE, NON-SPORE FORMING RODS (GROUP-16)

The main part of this group are the Lactobacillaceae organisms associated mainly with milk and milk products (G+C ratio = 35-53 mole %). The genera *Lactobacillus, Leuconostoc, Streptococcus, Pediococcus* and *Bifidobacterium* belong to this category. However, only the rod-shaped bacteria of the genus *Lactobacillus* are included in group-16 of Bergey's Manual. The coccoid genera are assigned to group-14, while *Bifidobacterium* is included among the actinomyces (Group-17). Besides lactic acid bacteria some pathogenic genera *Listeria* and *Erysipelothrix* are also included in this group. *Listeria monocytogenes* causes infection of warm blooded animals e.g., circling disease of cattle. *Erysipelothrix rhusiopathiae* causes 'wine erysepelas'.

4. CORYNEFORM BACTERIA (GROUP-17)

Coryneform bacteria (G+C ratio = 48-70 moles %) include a wide range of diversity covering (i) human and animal pathogens (e.g., *Corynebacterium diphtheriae, C. pyogenes*) (ii) plant pathogen (e.g., *C. fascians*) and (iii) non-pathogenic soil organisms including the genera of *Arthrobacter* and *Kurthia*.

Corynebacterium: (Greek *Coryne*, club). The genus comprises forms of the *Corynebacterium diphtheriae* type. These organisms are characterised by a snapping apart of the cells during cell division; the connecting walls on either side of the newly made cell wall appear to separate at different rates, so that the two cells seem to twist away in opposite directions. *C. diphtheriae* is the causative agent of diphtheria. It grows as a microaerophile or even anaerobes while most corynebacteria are aerobic. Its pathogenicity is due to its invasion on the larynx and tonsils and production of an exotoxin that circulates in the blood and attacks cells of the cardiac muscle, kidney and nervous system giving rise to post diphtherial paralysis. *C. mediolaneum* was the first bacterium used for biological conversion of steroid.

Arthrobacter :   An inhabitant of soil, *Arthrobacter* is characterised by a marked tendency of branching and having coccoid forms. Some members are motile and flagellated and all are aerobic. Another group of coryneform bacteria in the soil consists of several cellulose-utilising species of the genus *Cellulomonas*.

The coryneform bacteria can be regarded as an intermediate group between the lactic acid bacteria and mycobacteria in morphological and physiological properties. The tendency for branched cell formation increases in the series propionibacteria, coryneform bacteria and mycobacteria. In the same series there is a general transition from anaerobic to strictly anaerobic metabolism.

5. MYCOBACTERIA (GROUP-17)

Mycobacteria are invariably aerobic. Morphologically they are intermediate between the corynebacteria and proactinomycetes (*Nocardia*). They do not form mycelia but grow in the form of irregularly shaped branched cells. They are non motile and Gram-positive; they differ from corynebacteria in being acid fast.

Corynebacterium, *Mycobacterium* and *Nocardia* exhibit a number of similarities in their cell wall compositions. Their murein skeleton resembles that of Gram-negative bacteria but is complexed with an **arabinogalactan**, a polysaccharide consisting of arabinose and galactose. Some medically important members of this group are *M. tuberculosis, M. leprae, M. enteritidis*.

6. ACTINOMYCETES (GROUP-17)

The actinomycetes are a group of Gram-positive bacteria which produce very fine mycelium. They are aerobic with very few exceptions. The name of this group is derived from the first described anaerobic species *Actinomyces bovis* which causes **actinomycosis**, the 'ray-fungus' desease of cattle. Bergey's Manual of Determinative Bacteriology (8th edition) has grouped eight families within the order Actinomycetales. Members of the large genus *Streptomyces* have permanent mycelia; their aerial mycelia are often very well developed and contain aerial sporophores. These sporophores are often grouped to form structure resembling the coremia or pycnidia of true fungi. Knowledge about streptomycetes has advanced considerably because of their practical utility as producers of many effective antibiotics for example, Streptomycin form *S. griseus*, Chloramphenicol from *S. venezuelae* and Aureomycin and Tetracyclin from *S. aureofaciens*. Many streptomycetes degrade cullulose, chitin and other recalcitrant natural substances (e.g., *Micromonospora, Streptosporangium*). Some actinomycetes do not produce spores directly on the aerial mycelia but within sporangia (e.g., family Actinoplanaceae). A pathogenic species, *Dermatophilus congolensis* produces motile spores and causes dermatitis of the dorsal skin in sheep and horses.

7. ENDOSPORE FORMING RODS AND COCCI (GROUP-15)

These bacteria, in general, are grouped in the family Bacillaceae which have characteristic ability to produce more or less heat-resistant spores. The aerobic and facultative anaerobic rods belong to the genera *Bacillus* (G+C ratio = 36-62 mole %), *Sporolactobacillus* and *Sporosarcina* and the anaerobic rods belong to the genera *Clostridium* (G+C ratio = 23-43 mole %) and *Desulfotomaculum*. The two representative spore formers *Bacillus* and *Clostridium* contain organisms essential to soil fertility. The bacilli are usully differentiated according to the shape of their spores. For example, the spores of *B. megaterium* *B. cereus*,

B. subtilis, B. licheniformis, B. anthracis and *B. thuringiensis* are oval or cylindircal and not wider than the vegetative cell. In *B. polymyxa, B. macerans, B. tearothermophilus, B. circulans* the oval spores are wider than the vegetative cell while in *B. pasteurii* the spores are almost spherical and terminally distended. Among anaerobes, the genus *Clostridium* usually lacks cytochromes and catalase. A large number of substrates are readily fermented by clostridia including polysaccharides, protein, amino acid, purine (e.g., *C. butyricum, C. acetobutylicum, C. cellulosae-dissolvens*). *Clostridium tetani* and *C. botulinum* are well known human pathogens.

GRAM-NEGATIVE BACTERIA

8. *PSEUDOMONAS AND OTHER GRAM-NEGATIVE RODS (GROUP -7,12,3)*

The family Pseudomonadaceae consists of Gram-negative, polarly flagellated, straight or slightly curved rods that grow aerobically and are not spore formers. They are **chemo-organotrophs**, though some are facultatively **chemolithotrophs**. The genus *Pseudomonas* is the prototype of this family. Some pseudomonads are pathogenic to man and animals (e.g., *Ps. aeruginosa, Ps. mallei*); others are plant pathogens (e.g., *Ps. solanacearum*, G+C ratio = 58-78 moles %). The yellow pigmented plant pathogens of the pseudomonadaceae have been unified in the genus *Xanthomonas*. The yellow pigment is a polyene compound containing bromine. Genera that resemble the pseudomonads metabolically are *Alcaligens, Agrobacterium, Rhizobium* (G+C ratio = 59-66 mole %), the acetic acid bacteria *Acetobacter* and *Gluconobacter* and the free living nitrogen fixers *Azotobacter*, (G+C ratio = 53-70 mole %), *Beijerinckia* and *Derxia*.

Aerobic chemolitho-autotrophs (Group-12) are characterised by their ability to use inorganic ions or compounds as electron or hydrogen donors. Autotrophic bacteria belong to many genera including *Pseudomonas, Alcaligenes, Xanthomonas, Mycobacterium, Bacillus, Nocardia* but Group-12 of the Bergey's Manual (1974) includes only the bacteria that utilise ammonia and nitrate as well as iron and sulphur compounds e.g., *Nitrosomonas, Nitrobacter* (G+C ratio = 51-62 mole %), *Thiobacillus* (G+C ratio = 50-68 mole %).

Bacteria grouped together as sheathed bacteria (Group-3) show greater affinity to pseudomonads. These bacteria are usually encased in gelatinous sheaths and are found in both clean and polluted water in an activated sewage sludge. The best known filamentous Gram-negative form with polytrichously polar flagellation is *Sphaerotilus natans*, commonly referred to as 'sewage mould'. Many bacteria of this group are able to oxidise iron and manganese and were formerly referred to as 'ochre' bacteria (*Lepotothrix ochracea*).

9. *GRAM-NEGATIVE FACULTATIVE ANAEROBIC RODS (GROUP-8)*

The members of this taxonomic group are characterised by their fermentation products. These are small rods, motile by peritrichous flagella or non motile, non capsulated, not acid fast, catalase positive. Many of the representative members of this group e.g., *Escherichia, Salmonella, Shigella* are included in the family Enterobacteriaceae in Bergey's Manual (Greek-*enteron*, intestinal tract; G+C ratio = 30-59 mole %). *E. coli*, in all probability is the most extensively studied laboratory organism. Another family Vibrionaceae of this group includes *Vibrio cholerae*, the causal organism of cholera. *Erwinia* is a phytopathogenic form.

10. *GRAM NEGATIVE ANAEROBIC BACTERIA (GROUP-9)*

Species of the genus *Bacteroids* of family Bacteroidaceae (*B. fragilis, B. succinogenes*) belong to the dominant Gram-negative flora of human feces. Cells are uniformly rod-shaped or pleomorphic, nonmotile or motile with peritrichous flagella. They are chemo-organotrophic and are obligate anaerobes; G+C ratio = 40 mole %. *Fusobacterium* and *Leptotrichia* are the other genera of this family, the former deriving its name from its spindle like shape and the later being in the form of non-flagellated thread like upto 200 μm long. Among this group, genera of uncertain affiliation is *Desulfovibrio* - a sulphur-reducing organism found in brackish water.

11. *METHANOGENIC AND OTHER ARCHAEBACTERIA (GROUP-13)*

The methanogenic bacteria have been given a separate status in the 8th edition of Bergey's Manual as Group-13. However, for various reasons these are now included alongwith two other groups of archaebacteria. Archaebacteria are found in rather extreme habitats supposed to have prevailed in the earliest times of the earth's development i.e., archaic time. They include lithoautotrophic and heterotrophic aerobes or anaerobes that are distinguished in many ways from the eubacteria. According to present state of knowledge the archaebacteria are presumed to be the early deviants from the eubacteria owing to their simpler cell organisation. Most of them are probably the progenies of the primordial bacteria that had discovered the utilisation of the inorganic hydrogen donar (H_2) and acceptor (CO_2, sulphur) which were available in the earliest era of development. Archaebacteria can be subdivided into three groups; methanogenic bacteria, halophilic bacteria and thermoacidophilic bacteria.

Methanogenic bacteria

Important genera of this group are *Methanococcus vannielli* (cocci), *Methanobacterium formicicum* (rods), *Methanobrevibacter ruminantium* (short rods), *Methanospirillum hungatii* (spirilla) etc. these bacteria are extremly sensitive to oxygen and are killed on short exposure to the atmosphere. G+C ratio varies from 42-61 mole %. The ecology and metabolism of the methanogenic bacteria have been discussed elsewhere.

Halobacteria

The genera *Halobacterium* and *Halococcus* comprise extreme halophites. They are aerobes or heterotrophs and are mostly found in brakish water. They have the special properties of being able to utilise light energy for their metabolism.

Thermoacidophilic bacteria

This group includes autotrophs or heterotrophs acidophiles and neutrophiles that may be aerobes or anaerobes. *Sulfolobus acidocaldarius* and *Thermoplasma acidophilus* are some interesting members of this group.

CURVED RODS AND FLEXIBLE CELLS

12. *CURVED RODS : SPIRILLA AND VIBRIOS (GROUP-6)*

The bacteria of this group are placed in the family Spirillaceae which contains two distinct genera : *Spirillum* and *Campylobacter*. This group contains the rigidly helical single or grouped polar flagella which impel the cells with characteristic cork-screw motility. The genus *Spirillum* (G+C ratio = 38-65 mole %) comprises both free living aquatic organisms (e.g., *Sp. anulus*) and pathogens (e.g., *Sp. minor*-a natural parasite of rats that gives rise to 'rate-bite fever' in man). *Sp. volutans* is frequently found in pig manure and is characterised by its volutin content.

    Campylobacter foetus is one cause of venereal disease of cattle (G+C ratio = 30-35 mole %). Some species of *Campylobacter* are common in the oral flora of man.

    Bdellovibrio bacteriovorus, which is often to be parasited on other bacteria (G+C ratio = 43-50 mole%) is also included in this group. In contrast to bacteriophages which can multiply only in growing bacteria, *Bdellovibrio* can attack and lyse non growing cultures of moulty Gram-negative bacteria.

13. *SPIROCHAETES* (GROUP-5)

The bacteria grouped as Spirochaetes are included in the order Spirochaetales and further divided into five genera : *Spirochaeta, Cristispira, Treponema, Borrelia* and *Leptospira*. These are unicellular chemoheterotrophic bacteria of very characteristic shape. Like spirilla, they are helical but the cell body is extremely flexible. These are extra-ordinarily small that can pass through most bacteriological filters. The spirochaetes lack flagella but are able to move or glide over solid surface. Free living spirochaetes are commonly found in aquatic environment, others belong to the normal autochthonus microflora of animals. Few are pathogenic to man causing syphilis (*Treponema pallidum* G+C ratio = 32-50 mole %), relapsing fever (*Borrelia recurrentis*), leptospirosis or Weil's disease (*Leptospira icterohaemorrhagiae*). *Spirochaeta plicatlis* is widely distributed in fresh water and is recognised by characteristic restless movement. *Cristispira* lives in the 'crystal stalk' and gut of fresh water and marine mussels.

LARGE SPECIAL GROUPS

GLIDING BACTERIA (GROUP-2)

Only few bacteria are able to move by gliding or creeping. These can be grouped as follows:

(i) Bacteria that contain intracellular sulphur of which there are **trichome** forming (*Beggiatoa, Thiothrix*) and unicellular (*Achromatium*) representatives.

(ii) Sulphur free bacteria existing as trichomes (*Vitreoscilla, Leucothrix, Saprospira*).

(iii) Unicellular rod-shaped bacteria including the myxobacteria (*Cytophaga* group and *Flexibacter* group).

(iv) The thread like gliding bacterium (*Chloroflexus*).

(v) Cyanobacteria, if motile, move by gliding.

Beggiatoa is a colourless, aerobic, thread like sulphur bacterium which looks like masses of spider's web. It consists of **trichomes** of uniform thickness. The cells are usually filled with sulphur droplets. The threads are motile by gliding. *Thiothrix* is not freely motile. Bunches or tufts of trichomes are attached by their bases to solid surfaces. Multiplication is by gonidia which are formed by the rounding off of apical cell. These gonidia can glide over solid surfaces. *Vitreoscilla* is a colourless, aerobic, multicellular, filamentous bacterium which moves by gliding and multiplies by fragmentation of the filaments. *Leucothrix* grows epiphytically on marine algae. Myxobacteria (G+C ratio = 67-70 mole %) are strictly aerobic chemoheterotrophic organisms with gliding motility. On soil, myxobacteria form very extensive flat colonies. Most of the myxobacteria produce battery of hydrolytic enzymes capable of lysing other prokaryotic or eukaryotic cells. The cytophaga group containing the genera *Cytophaga* and *Sporocytophaga* are known for their ability to degrade cellulolytic species. The *Flexibacter* species are aquatic bacteria. They contain carotenoid and show yellow, pink or orange pigmentation

15. *PROSTHECATE AND BUDDING BACTERIA (GROUP-4)*

The organisms of this group differ markedly from that of typical bacteria either by the formation of **prostheca** or buds or hypha-like protrusion or by the formation of stalk of holdfast (Fig. 4.1). A prostheca is a semi-rigid appendage, extending from a prokaryotic cell. Budding or sprouting is the term applied to the mode of multiplication as in budding yeast. Contrary to binary fission it is an unequal cell division. The daughter cell (bud) is usually smaller than the mother cell and reaches normal size only after it has separated from the mother cell.

The genus *Hyphomicrobium* is a typical example of prosthecate bacteria in which the prosthecae have a reproductive function (i.e., new cell formation by budding). *H. vulgare*, a denitrifying bacterium, is a regular inhabitant of stagnant water. Another organism with similar habits is the non sulphur purple bacterium *Rhodomicrobium vanniellii*. Bacteria of the genus *Caulobacter* unlike *Hyphomicrobium* do not bud from the apex of the stalk rather divide by binary fission of the cell proper. They have the distinctive

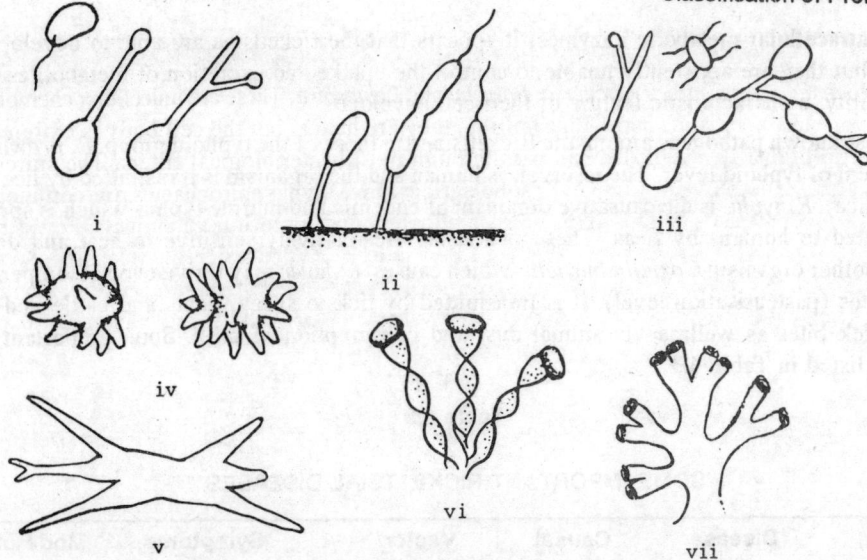

Fig. 4.1   Some prosthecate and stalked bacteria

i. *Hypomicrobium* ii. *Rhodomicrobium* iii. *Caulobacter* iv. *Prosthecomicrobium*
v. *Ancalomicrobium* vi. *Gallionella* vii. *Nevskia.*

feature of producing a very tine stalk which may be attached with the help of holdfast to form rosette or to some other substrates.

The genus *Gallionella*, the best known iron bacterium, grows profusely in the spring and in iron-containing waters and is characterised by forming appendages (stalk) and reproducing by binary fission. *G. ferruginea* is a bean shaped cell which excretes a slime on its concave side. Because of their ability to transform reduced iron to insoluble ferric compounds they may cause problems by clogging in water system.

16. *OBLIGATE CELLULAR PARASITES (GROUP-18)*

Some bacteria are strictly obligate that occur as endosymbionts in insects and ciliates. Rickettsiae and Chlamydiae are the best studied examples that are known as potent pathogens of animals and human beings.

RICKETTSIAE

Rickettsiae are obligate intracellular parasites of arthropods like fleas, lice, mites, ticks etc., in which they exist as harmless parasites or even symbionts and often are pathogenic to man. The name of this organism was given in 1916 to honour H.T. Ricketts, who discovered American **'Rocky Mountain Spotted fever'**. This group is also referred to as the spotted fever (typhoid fever) group on account of diseases produced by *Rickettsia prowazekii* which was first reported by S. Von Prowazek, another pioneer in typhus investigation. The disease holds its own significance in having long association with human sufferings and miseries during many historical wars and is regarded dreadful, only next to malaria and bubonic plague.

Rickettsiae resemble other bacteria morphologically occuring as short rods, cocci, in chains or filaments and as pleomorphic coccobacillary forms. They are non motile and non-spore forming. Like other Gram-negative bacteria their cell walls contain muramic acid. They stain weakly with aniline dyes and are lysozyme sensitive. Although the size of Rickettsia is equivalent to that of pox-virus, they can be differentiated unequivocally from any virus in having DNA as well as RNA in a ratio of 1:3.5 and possessing intracellular metabolic enzymes. It appears that the rickettsiae are able to develop their own metabolism but they are apparently unable to control the uptake and excretion of metabolites because of the permeability, a characteristic feature of their cell envelope.

The best known pathogens among the Rickettsiae are those of the typhoid group. *R. prowazekii* is the causative agent of typhoid fever. The reservoir is human and the organism is transmitted by lice (head lice and clothes lice). *R. typhi* is the causative organism of endemic and murine typhus which is spread by rats and transmitted to humans by fleas. These rickettsiae are relatively sensitive to heat and dehydration. However, another organism *Coxiella burnetii*, which causes *Q-fever* may thrive even at temperature 60°C for 30 minutes (pasteurisation level). It is transmitted by tick to sheep, goats and cattle and can infect human by tick bites as well as via animal dust and consumption of milk. Some important rickettsial diseases are listed in Table 4.3.

Table 4.3

SOME IMPORTANT RICKETTSIAL DISEASES

Biotype	Disease	Causal Organism	Vector/ (Reservoir)	Symptoms	Mode of Transmission
1. Typhus group	Epidemic typhus	*R. prowazekii*	Human body louse (Human)	Fever	Through the skin infected flea feces
2. Typhus group	Endemic typhus	*R. typhi*	Rat, flea & Louse	Fever & scrotal swelling	DO
3. Spotted fever group	Rocky Mountain spotted fever	*R. rickettsii*	Ticks (wild rabbits, dogs, sheep, rodents)	Fever & scrotal necrosis	Tick bites
4. Spotted fever group	Rickettsial pox	*R. akari*	House mites (mice)	Fever & scrotal swelling	Mite bites
5. Spotted fever group	Siberian tick typhus	*R. sibirica*	Tick (rodents)	DO	Tick bites
6. Spotted fever group	Trench fever	*R. quintana*	Human body louse (Human)	--	Tick bites
7. Spotted fever	Scrub typhus	*R. tsutsugamushi*	Mites (rodents)	--	Mite bites
8. Q fever group	Q fever	*Coxiella burnetii*	Tick and body lice (rodents, cattle, sheep, goat)	--	Inhalation of aerosoles, Tick-bites, milk

CHLAMYDIAE

The genus name *Chlamydia* is derived from the Greek *Chlamydion*, meaning 'small cloak or mantle'. Only two species are recognised. *C. trachomatis* causes trachoma, the Egyptian eye disease that starts as **conjuctivitis** and leads to blindness and the venereal disease **lymphogranuloma-venereum**. In both the cases the disease is transmitted by contact. *Chlamydia psittaci* is the causative agent of **ornithoses**, the best known of which is psittacosis, a feverish pneumonia. Birds are the main hosts of chlamydiae.

Chlamydiae are Gram-negative, basophilic, coccoid or spherical cells. They contain RNA and DNA in a ratio characteristic of bacteria and they synthesize substances which eukaryotic cells are unable to produce such as muramic acid, diaminopimelic acid, D-alanine, folic acid etc. The outstanding feature that distinguishes chlamydiae from rickettsiae is their dependence on the host cell for their energy. They are unable to phosphorylate glucose or metabolise it. They are, however, extra-ordinarily permeable to ATP and CoA. For this reason chlamydiae are regarded as 'energy parasite'.

17. MYCOPLASMA (GROUP-19)

The members of Mycoplasma group are the smallest independently replicating prokaryotes. The organisms lack a true cell wall and have probably been evolved from a number of different origins. Because of their apparent role in pathogenesis the organisms have received considerable attention in the recent years. These have been described in details in Chapter-V.

18. ANAEROBIC ANOXYGENIC PHOTOTROPHIC BACTERIA (GROUP-1)

Two groups of bacteria have the ability to use light as energy source for growth : the anaerobic phototrophic bacteria and the aerobic oxygenic cyanobacteria. The former belonging to order Rhodospirillales are characterised by their possession of photosynthetic pigments (bacteriochlorophyll and certain carotenoids). These bacteria exhibit their dependence on light for energy but are not able to utilise water as the hydrogen donor. They are obliged to use more reduced hydrogen donors (H_2S, H_2 or organic compounds). These are mostly predominant in aquatic environment having spherical, rod, vibrio or spiral cell morphology. Four families are distinguished on the basis of important physiological properties: (i) Purple sulphur bacteria (Chromatiaceae). (ii) Non sulphur purple bacteria (Rhodospirillaceae) (iii) Green sulphur bacteria (Chlorobiaceae) and (iv) Chloroflexus group (Chloroflexaceae). The majority of the purple sulphur bacteria are easily recognised by their intracellular presence of highly refractile sulphur globules e.g., *Thiospirillum jenense, Chromatium warmingii*. The non sulphur purple bacteria (Rhodospirillaceae) are usually represented by two genera viz; *Rhodopseudomonas (R. palustris, R. viridis, R. acidophila)* and *Rhodospirillum (R. rubrum, R. tenue, R. fulvum)*. The green bacteria (Chlorobiaceae) comprise green (*Chlorobium vibrioforme, C. limicola*) as well as brown pigmented strains (*Chlorobium phaeobacteroids*), starshaped aggregations (*Prosthecochloris*) and net forming organisms (*Pelodictyon clathratiforme*). The phototrophic green bacterium *Chloroflexus* belongs to the filamentous gliding bacteria according to its shape and type of motility. It contains bacteriochlorophyll *c* and *a*. *Chloroflexus* differs from *Chlorobium* species by its capacity for aerobic, heterotrophic growth in complex media in the dark as well as in light.

19. AEROBIC OXYGENIC PHOTOTROPHIC BACTERIA CYANOBACTERIA

In contrast to anaerobic phototrophic bacteria, cyanobacteria do utilise water as hydrogen donor and evolve oxygen in the light i.e., they carry out oxygenic photosynthesis. The pigment system includes chlorophyll *a*, carotenoids and phycobilins. The cyanobacteria have been described in detail in Chapter-II and will not be discussed here (Fig. 4.2).

Recently a new photosynthetic organism that has serveral characteristics common to prokaryotes and green algae has been discovered. It has the typical cell structure of prokaryotes (Gram-negative cell wall with muramic acid, no organelles or true nucleus) but it contains in addition to chlorophyll *a*, chlorophyll *b* also which is found in any green algae. It also differs from cyanobacteria by the absence of phycobilin protein, cyanophycin and poly *B*-hydrobutyrate. This transitional form has been called **Prochloron**.

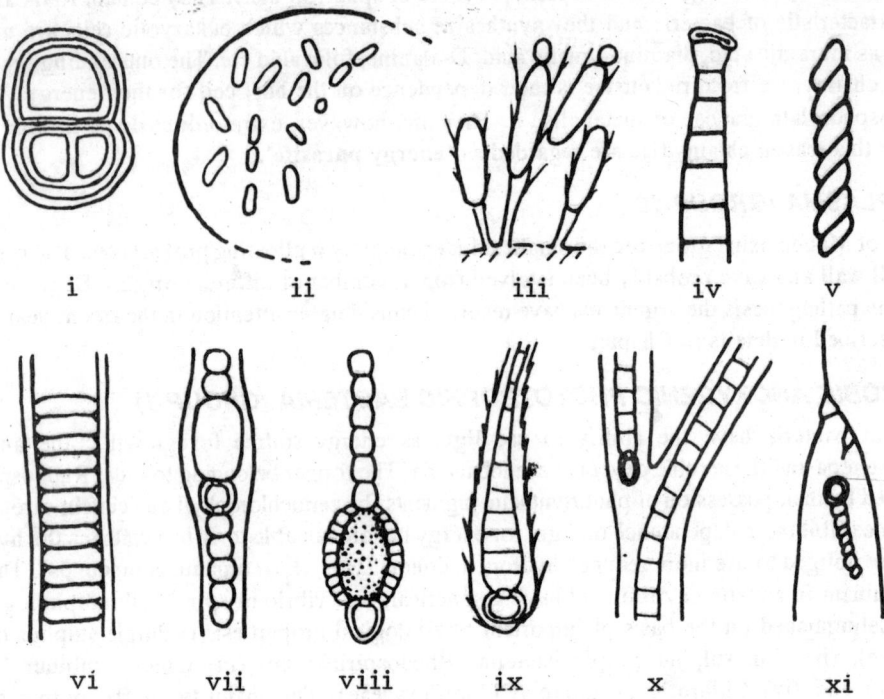

Fig. 4.2   Some Cyanobacteria

i. *Chrococcus* ii. *Aphanotheca* iii. *Chamaesiphon* iv. *Oscillatoria* v. *Spirulina*
vi. *Lyngbya* vii. *Anabaena* viii. *Cylindrospermum* ix. *Calothrix* x. *Tolypothrix*
xi. *Richelia.*

SUGGESTED READINGS

Buchanan, R.E. and V.E. Gibbons (1974) (eds). Bergey's Maunal of Determinative Bacteriology, 8th ed. Williams and Wilkins, Baltimore.

Krieg, N.R. and Holt, J.G. (1984) Bergey's Manual of Systematic Bacteriology, Vol. I Williams and Wilkins, Baltimore.

Lapage, S.P., Sneath, P.A.M., Lessel, E.F. Jr., Skermann, V.B.D., Seelinger, H.P.R. and Clark, W.A. (1975). International Code of Nomenclature of Bacteria. American Society for Microbiology, Washington D.C.

Laskin, A.I. and H.A. Lechevalier (eds) (1974). CRC Handbook of Microbiology. CRC Press, Inc: Cleveland.

Schleifer, K.H. and Stackebraudt, E. (1983). Molecular Systematics of Porkaryotes. Ann. Rev. Microbiol. **37**: 143.

Sneath, P.H.A. and Sokal, R.R. (1973). Numerical taxonomy : The Principle and Practice of Numerical classification. Freeman. San Francisco.

Starr, M.P., Stolp, H., Truper, H. G., Balows, A. and Schlegel, H.G. (eds) (1981). The Prokaryotes : A Handbook on Habits, Isolation and Identification of Bacteria, Vol. I and II. Springer, Heidelberg.

5

MYCOPLASMAS

In the early eighteenth century bovine pleuropneumonia appeared as a distinctly recognisable contagious disease in Europe. For many years the infectious agent was thought to be a virus. Repeated attempts to cultivate the agent on the common bacteriological culture media known till that time met with disappointment. Credit should go to Nocard, Roux and collaborators who described the first isolation of the prototype of a group of micro-organisms in 1898 now known as **Mycoplasma**.

Spurred on by this initial success other workers (Dujardin-Beaumetz, 1906; Borrel *et al*, 1910; Bordet, 1910) studied the morphology and infectivity of this new microbial agent. Bordet descirbed the variable morphology stating that in some cases it looked like the virus of syphilis and that it could be stained with Giemsa stain. Borrel *et al* (1910) observed the characteristic pleomorphism describing asterococcal, round and ovoid granular, tetrad ring, pseudovibrio and filamentous forms. The lag phase of studies with mycoplasmas continued through the 1930's and 1940's. Morphological examination of the organism of bovine pleuropneumonia was extended by Turner (1933, 1935) and Orskov (1939). Organisms with cellular and colonial morphology of the bovine pleuropneumonia organism, were sought and found in a variety of sources. The lack of an acceptable classification scheme led to their being designated as **pleuro-pneumonia like organisms** or PPLO. Shoetensak (1936) successfully recovered such organisms from dogs suffering from distemper.

The first isolation of mycoplasmas from the human was made by Dienes and Edsall (1937) who found it as the apparant cause for suppuration of the Bartholin's gland. Subsequently mycoplasmas were found in the genitourinary tract of humans by many workers. The role of these organisms in producing diseases was unequivocal. *Mycoplasma pneumoniae* was successfully reported to be the cause of a typical pneumonia in human by Eaton *et al* (1944). Other infections of humans were shown to be associated with mycoplasmas in the late 1940's and early 1950's. Some important associations of mycoplasmas in humans are presented in Table-5.1.

Detection of mycoplasmas was not restricted to animals alone. In plants also mycoplasma like organisms were found to be associated with many diseases. **Paulownias**, popularly known as 'Empress tree' or 'Princess tree' belonging to family Bignoniaceae, grows in China (Taiwan), Korea and Japan. For the first time **witche's broom** of paulownias was observed in Japan in 1880 (Kawakami, 1902). Since then more than 80 plant diseases which were ascribed earlier to be due to virus causing witche's broom symptoms and which were designated as yellow type of diseases have been known to be associated with mycoplasma like organisms.

Table 5.1

MYCOPLASMAS ASSOCIATED WITH HUMAN BEINGS

Species	Site of isolation	Nature of disease
M. orale	Oropharynx	Non pathogenic
M. salivarium	Oropharynx	Non pathogenic
M. buccale	Oropharynx	Non pathogenic
M. faucium	Oropharynx	Non pathogenic
M. laidlawii	Oropharynx	Non pathogenic
M. fermentans	Genitourinary tract	Non pathogenic
M. primatum	Genitourinary tract	Non pathogenic
M. hominis	Genitourinary tract	Septicemia, abscess, endometritis - salpingitis, reproductive disorders (sterility, abortion)
M. urealyticum	Genitourinary tract	Non gonococcal urethritis, septicemia, reproductive disorders.
M. pneumoniae	Oropharynx, respiratory tract	Tracheo-bronchitis pneumonia, hemorrhagic bullous myringits.

Japanese workers (Doi *et al*, 1967) described MLO (Mycoplasma Like Organisms) as plant pathogens while investigating the cause of mulberry dwarf disease. Since then work on mycoplasma and their association with plant diseases have been carried out in many countries including India. A good deal of work on Aster yellows, which was extensively studied by Kunkel and his associates since 1923, was eventually proved by Maramorosch *et al* (1972) to be due to **Spiroplasma**. Edward and Freundt in 1956 proposed to retain the generic name Mycoplasma for "Pleuropneumonia like Organisms" or PPLO. The same authors in 1967 separated mycoplasmas from the class Schizomycetes and proposed a new class **Mollicute** which is now universally accepted (Table-5.2).

Table 5.2
CLASSIFICATION OF MOLLICUTES

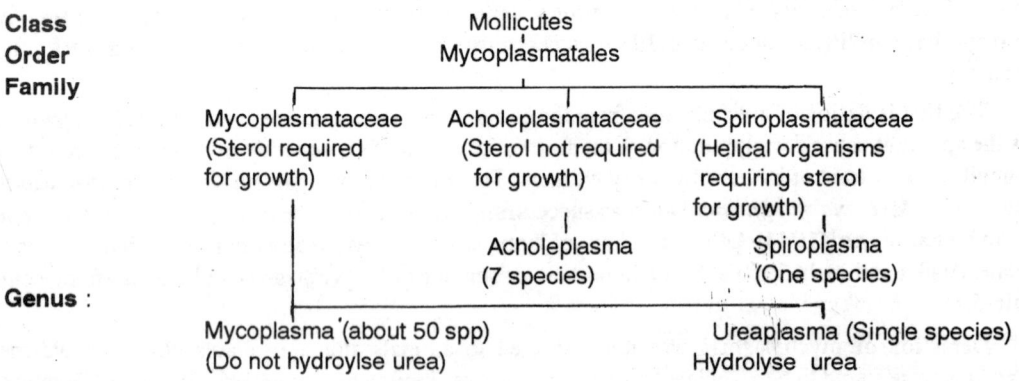

Class	Mollicutes
Order	Mycoplasmatales
Family	Mycoplasmataceae (Sterol required for growth) / Acholeplasmataceae (Sterol not required for growth) / Spiroplasmataceae (Helical organisms requiring sterol for growth)

Acholeplasma (7 species)   Spiroplasma (One species)

Genus :   Mycoplasma (about 50 spp) (Do not hydroylse urea)   Ureaplasma (Single species) Hydrolyse urea

MYCOPLASMA - A SEPARATE GROUP

The organisms of the pleuropneumonia group are classified under the proposed new class Mollicutes (L. *Mollis*, soft or pliable; *Cutis*, skin). Under this class the order Mycoplasmatales or alternatively Mollicutales nas been established. The former term derived from the Greek noun *myces* (meaning fungus) and *plasma* (meaning something formed or molded) has gained common acceptance.

Granting an independent status to *Mycoplasma* within Bergey's Manual of Determinative Bacteriology had been a debatable topic. Many workers have tried to trace its evolutionary lines from bacterial stock, some have shown its close proximity to bacterial **L-phase** and some have gone to the extent of finding a link with viruses. However, all workers in this field acknowledge the value of an orderly nomenclature as contrasted to symbolism. A comparative study of bacteria, mycoplasmas and viruses has been presented in Table-5.3. The fundamental question requiring an answer is whether sufficient distinction exists between these organisms and other related forms. Scrutiny of these points deserves mention.

Table 5.3

A COMPARATIVE ACCOUNT OF MYCOPLASMAS, BACTERIA AND VIRUSES

Properties	Mycoplasmas	Bacteria	Viruses
Size (diam.)	$0.3\,\mu m$	$1\text{-}2\,\mu m$	$0.5\,\mu m$
Lack a cell wall	Yes	No	Yes
Propagate on cell-free media	Yes	Yes	No
Require sterol and native protein for growth	Yes	No	No
Intrinsic energy metabolism	Yes	Yes	No
Narrow range of host specificity	Yes	No	Yes
Growth inhibited by specific antibody	Yes	No	Yes
Resistant to penicilin	Yes	No	Yes
Resistant to antibiotics that inhibit metabolism (e.g., Tetracyclin)	No	No	Yes/No

Diversity of the organism

Mycoplasmas exhibit high degree of diversity with respect to their nutritional requirement, metabolic activities, DNA composition, protein components and ubiquity of their occurrence which support their existence as a separate class. They occur as parasite or commensals in most mammalian species. The guanine plus cytosine content varies to the range of 23 to 39% suggesting to be simply one genus.

Morphological evidence

Colonial morphology is distinctive. The colonies which vary from 10 to $500\,\mu m$ in diameter typically present an **umbonate shape** i.e., the appearance of a **fried egg**. The central portion is denser as compared to the peripheral area. Sometimes the colonies exhibit a lacy net work. No true bacterial colony displays this form and size. Electron micrographs of thin sections clearly show that the limiting envelope is a trilaminar unit membrane and no vestige of an outer rigid wall exists. The resultant plasticity of the organisms undoubtedly represents the deceptive morphology of individual cell.

Earlier some morphologists alluded to the similarities of mycoplasma to viruses. Their filterability led them considered viruses for many years. Pleomorphism during growth and development suggested a possible relationship to infectious ectromelia virus of mice. The trilaminar envelope enclosing DNA strands and ribosome is also met in ornithosis virus. Recently morphological examination by electron microscopy clearly shows distinction from the rickettsiae, which possess a cell wall.

Nutritional requirement

Some nutritional requirements and many growth conditions like pH, temperature, gaseous environment etc. have served as useful parameters in distinguishing one species of *Mycoplasma* from another. Need of some sterols is the uniqueness of mycoplasmas. However, this characteristic is not always reliable. There are some species of *Mycoplasma* (*M. laidlawii*) that possess no requirement of sterol.

Metabolic activities

The mycoplasmas as a group display a wide diversity of metabolic activities. Many degrade sugars with the production of lactic acid and acetic acid similar to lactic acid bacteria. Some oxidise short chain fatty acids by the B- oxidative pathway. Others convert arginine to ornithin by the *arginine desamidase* pathway. A special group known as T strains hydrolyses urea. However, no common metabolic character has so far been found that can distinguish mycoplasmas from other micro-organisms. A peculiar property associated primarily with mycoplasmas is the inhibition of growth by specific antisera. Whether the metabolic inhibition reflects interference with the specific enzymes involved in substrate utilisation or whether some other general phenomenon occurs has not yet been established.

DNA compositions and Homology

The wide variation in base composition of DNA suggests that mycoplasmas are heterogenous. DNA-DNA and DNA-RNA homology studies have revealed striking heterogenecity among the mycoplasmas. Moreover no significant homology has been detected between mycoplasmas and bacteria and the homology data support the classification of mycoplasmas as distinct and different from the bacteria.

Protein composition

Electrophoretic pattern of the total cellular or membrane proteins have proven useful in differentiating species of mycoplasmas. Each species gives a distinctive pattern of protein profile unlike the bacteria. However, there is paucity of such comparative informations on bacteria and mycoplsmas. Investigations on this aspect may be rewarding and useful.

Serological characteristics

A large number of attempts to demonstrate meaningful serological cross relationship between bacteria and mycoplasmas have not met with success so far. Freundt (1958) using *Streptobacillus moniliformis* and various mycoplasmas as well as Shifrine and Gourlay (1967) using *M. mycoides* and ten different bacterial species did not observe any positive relationship in their experiments. This is possibly on account of fact that bacteria possess outer walls which constitute much of their dry weight and their antigenic characters. Removal of the walls results not only in loss of antigens but also in reduction of the specificity of antigens. Mycoplasmas which have no cell walls exhibit less immunological specificity as well as poorer antigenicity.

Isolation of bacteria from mycoplasmal cultures

The old bacterioloigcal literature is filled with claims and counter-claims of filterable forms which give rise to typical bacteria. There are reports that mycoplasmal cultures have yielded *Haemophilus* diphtheroids and Gram-positive cocci. However, these reports can not be considered authentic as these could not be reproduced in subsequent experiments. No proof could ever be furnished that bacteria did arise from the mycoplasma. And the reports could be taken as mere cases of contamination of mixed cultures due to experimental laxity.

The above evidences unequivocally emphasise that mycoplasmas are distinct from the bacteria. Yet no objective scientist can dismiss the possibility that some sort of relationship does exist which will require future resolution. Diversity of organisms, nutritional requirements and metabolic characteristics are too broad to be useful in distinguishing mycoplasmas as a group separate from other micro-organisms. The greatest similarity of mycoplasma, if one were to choose, lies with the Gram positive bacteria, in particular with the streptococci. Here is a resemblance in their complex nutritional requirements, metabolic and respiratory activities, specific glycolipids and to some extent in the base composition of DNA. Interestingly the demarcating line between bacteria and mycoplasma is not much distinct as there exists the L-phase of bacteria, these wall less bacteria capable of independent reproduction and growth, showing remarkable resemblance to mycoplasmas at one hand and bacterial parentage on the other.

STRUCTURE OF THE MYCOPLASMAL CELL

Colonial appearance

The typical mycoplasmal colony is round with a well demarcated edge. Its average diameter is about 100μm but may vary from 10 to 600 μm. It exhibits a dense centre and a translucent periphery giving the so called fried-egg shape appearance. The colonial appearance sometimes is characteristic of species or strain. For example the T-strains have very small colonies about 10 μm in diameter and lack peripheral surface growth. In species inhabiting humans, the diameter of the dense centre varies considerably. The periphery may appear granular or lacy. The lace-work has been attributed to lipid accumulation, or to vacuolisation as a result of lysis, or to the large flat cells containing condensed granules around the periphery. Increased vacuolisation is a characteristic of aging colonies.

Cellular morphology

One feature common to all mycoplasmas is the plasticity brought about by the absence of a rigid outer wall. Due to this reason all sorts of shapes ranging from coccobacilli to asteroidal structure are seen in mycoplasmas. However, cell size of the organism has still been a subject of criticism as the organism could change shape and pass through filters of pore size smaller than the actual organism. Nevertheless, in order to justify some generalisation, detailed studies have been performed with three species viz., *M. mycoides, M. laidlawii* and *M. gallisepticum*. The cell of *M. mycoides* is coccoid shaped measuring 250 to 300 μm. Very thin optically homogenous filaments are extruded from the cell which terminate in a tiny refractile spherical or club-shaped body. These filaments can be exceedingly long and no transverse septa are discernible in this filaments. The morphological unit of *M. laidlawii* is also a coccoid element about 500 nm in diameter. The cqccoidal or ellipsoidal forms are interspersed with smaller spheroidal elements avereging 125 nm. Colonies of *M. gallisepticum* possess polygonal and tear-drop shaped cells, 500 to 800 nm in diameter. The cells possess a knob-like protrusion at one or both ends. The morphological feature of *M. gallisepticum* appear more constant and well defined than in other mycoplasmas.

Chemical composition of the cells

The major differences of mycoplasmas from bacteria are the increased lipid content and the absence of wall components. The cells are composed of 40 to 60 % protein based on dry weight. Hydrolyzates of whole cells contain at least seventeen different amino acids, common to proteins. There is a complete absence of diaminopimelic acid. Starch gel and polyacrylamide-gel electrophoresis have shown a great mulitiplicity of protein bands which exhibit similar or dissimilar mobilities among different species. Carbohydrate contributes little to the total dry weight amounting to 0.1% or less. The DNA content ranges from 1.5 to 7% of the dry weight. No odd base like hydroxymethyl cytosine is found. RNA content varies from 3 to 17%. Lipid comprises 8 to 20 %, the lower values being characteristic of sterol non-requiring species. About half of the lipids are neutral comprising unsaponifiable lipids, glycolipids and some glycerids; the other half are polar lipids, exclusively phospholipids. The lipid phosphorus content ranges from 0.12 to 0.44%.

ULTRASTRUCTURE

The fine structure of a typical mycoplasmal cell consists of a trilaminar membrane surrounding a cytoplasm packed with ribosomes, fibrillar DNA, one or more electron dense areas and occasionally empty vesicles surrounded by a trilaminar membrane.

The cytoplasmic membrane is typical of the so called **unit membrane** being composed of a light area about 5 nm thick bounded on either side by electron dense region about 3 nm thick giving the membrane overall thickness of 11 nm. The nuclear material consists of a an unbounded fibrillar and granular region. The fibrils are about 3 nm thick and are not contained within a membrane reminiscent of the bacteria. The chrososome is circular. The size of the chromosome is approximately 1000×10^6 dalton allowing for coding of more than thousand cistrons. Ribosomes measure about 14 nm in diameter. Sedimentation analyses show a single component with a sedimentation coefficient of 72 S which on dilution with water yields three fractions with 'S' values of 70, 49 and 32, similar to bacterial ribosome.

The dense regions which occur as protrusion are termed **'blebs'**. The ultrastructure of bleb consists of a dense elliptical outer plate adjacent to the cytoplasmic membrane, a flat plate and fine threads connecting the flat plate to elliptical plate. Behind the bleb lies an **infra-bleb** region which is considered to be protein. Ribosomes are noticeably absent in this area, yet these blebs are seen in actively dividing cells.

Electron micrographs of thin section of membrane of some species of *Mycoplasma* appear to possess some cross striations. Besides, dense cytoplasmic bodies have been observed in most of the species. These dense regions are supposed to be elementary bodies that extrude from the cells. These are considered to be the sites of reproduction. Another structure frequently seen in majority of the species of *Mycoplasma* are empty vesicles surrounded by a trilaminar membrane. These structures are more prevalent in aging cells. It is also conceivable that they represent vacuoles from **pinocytosis** or mesosomes which are common to bacteria.

Motility has been observed in a strain of *M. pulmonis*. Motion was characterised by gliding rod-shaped forms and spinning spherical forms. Electron micrographs of intact organisms failed to reveal any flagellar structure.

CHEMICAL COMPOSITION OF STRUCTURAL UNIT

Extracellular polysaccharide

The amorphous extracellular material is a galactan. Acid hydrolysates contain more than 90% galactose and only a trace of glucose. It has been identified as polysaccharide with the predominance of *B*-linkage, 6-O-*B*-D -galactofuranosyl-1 (Fig. 5.1). It possesses serological activity in precipitin and indirect hemagglutination in reactions.

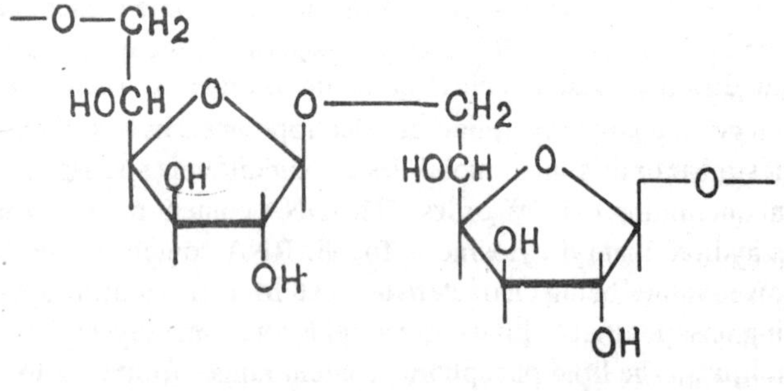

Fig. 5.1 Structure of main repeating units in polysaccharides
of *M. mycoides.*

Cytoplasmic membrane

The lining membrane of mycoplasmas is lipoprotein in nature. The overall chemical composition of the cytoplasmic membrane shows a protein content of 47 to 60% of dry weight, 35 to 37% lipid, 4 to 7% carbohydrate, 1 to 4% RNA and 1 to 2% DNA. The nucleic acids associated with the membrane may represent attached or trapped ribosomes and fragments of genome. Whether these components are integral parts of membrane is still unsettled. The carbohydrate reflects primarily the sugars found in glycolipids. A portion of it is associated for hexosamine which occurs in the membrane in variable amount. A limited number of enzymes have been found associated with the cytoplasmic membrane. These include glucosidase, reduced nicotinamide adenine dinucleotide oxidase, adenosine triphosphatase, ribonuclease, deoxyribonuclease, cholesterol esterase and phosphatidyl glucose synthetase. The nature of lipid varies in different species. However, a generalisation can be made that all mycoplasmas contain sterols or carotenols, glycerophospholipids and in many cases glycolipids.

Nuclear material

The DNA of mycoplasmas contain the usual bases - adenine, guanine, cytosine and thymine - but no hydroxymethyl cytosine. Adenine to thymine and guanine to cytosine ratio approximate 1.0 in agreement with the Watson-Crick model.

Ribosomes

No chemical analyses on purified ribosomes of mycoplasmas have been reported. The base composition of total RNA has been performed on *M. arthritides* and *M. gallisepticum*. The purine/pyrimidine ratio in *M. arthritides* is 0.98 whereas in *M. gallisepticum* it is 1.3. No obvious relationship between the base ratio of DNA and RNA is seen. The transfer RNA of mycoplasmas undergoes sedimentation with *E. coli* RNA. The nucleotide content of tRNA of mycoplasmas is lower than that of other microorganisms. Most of the species of mycoplasma contain N-formylmethionyl tRNA.

Bleb and Infra-bleb region

These ultrastructures have not been isolated for chemical analysis. Histochemical techniques suggest that they are protein possibly interspersed with lipid. There is notable absence of nucleic acid in this region. In *M. gallisepticum* ATPase activity appears to be localised in the bleb and infra-bleb regions exclusively. Acid phosphatase is localised only in infra-bleb region and does not appear to be membrane associated.

PHYSICAL NATURE OF STRUCTURAL UNITS

Cytoplamic membrane

The mycoplasmal membrane possesses the morphological and compositional features similar to most biological membrane. Besides, mycoplasmas have one additional feature, the presence of only one membrane structure. This has generated interest in using membrane for models to study the structure of membrane in general. The organism, when grown in the absence of cholesterol lyses readily in the medium of low ionic strength. The resultant membrane when thoroughly washed and treated with RNase and DNase appear as collapsed empty sacs made of lipoprotein. The cellular contents appear to have leaked from large holes in the membrane. Washing of intact mycoplasmas in hypotonic or hypertonic solution results in leakage of ultraviolet absorbing intracellular components. Uranyl ions prevent sterol incorporation into mycoplasmas and protects against lysis by detergent. Pancreatic lipase induces lysis of intact mycoplasmas. Phospholipase causes lysis of cells and solubilises purified membrane. These findings suggest a role for phospholipids and glycolipids in maintenance of structural integrity of the cytoplasmic membrane.

Chromosome

The chromosomes are circles of typical double stranded DNA as revealed by direct electron microscopy. Some species e.g., *M. arthritides* show a number of small circles which may be artefacts of breakage or *episome*. These small circles of DNA may represent the satellite DNA. All species examined so far, exhibit double 'Y's or forks confirming a typical replicative mechanism.

Ribosome and mRNA

M. gallisepticum is the best studied organism with respect to the physical characteristics of RNA. The ribosomal fraction contains a single component with a sedimentation coefficient of 72.2 ± 6.5 S. Ribosomes have a diameter of about 14 nm and are loosely arranged in young cells. As the cells grow, they assume a cylindrical packing arrangements. In either form they appear to concentrate near the cytoplasmic membrane. About 5% of the total RNA is messenger RNA which has a half life of about two minutes.

MODELS OF STRUCTURAL UNITS

Two different basic models for membrane structure have been advanced. One is the **lamellar model**, first proposed by Danielli and Davson (1935) and modified as the unit membrane by Robertson (1958). The other is the **corpuscular model** proposed by Green and Perdue (1966) and modified as sub-unit model by Lenard and Singer (1966).

The unit membrane model is visualised as a bimolecular leaflet of phospholipids held together by hydrophobic interactions of the hydrocarbon chains and sandwitched between two layers of protein. The lipid and the protein layers are considered separate but continuous phases. The proteins are considered to be globular. Robertson describes two protein layers to be asymmetric, the outer layer being mucopolysaccharide or mucoprotein and the inner layer to be unconjugated protein. The protein layers are held to the inner lipid leaflet by electrostatic forces between polar groups on lipid and protein. The protein layers are visualised as extended polypeptide chains about 2 nm thick and the middle layer 3.5 to 4 nm thick. The lamellar structure is shown in fig.5.2. Much of the evidence supporting this model is biophysical that includes the ubiquitous presence of lipid in all membranes, the electron transparency of the middle layer, the eletron density of the outer layer and the dimension of these layers. However, proponents of the corpuscular model counteract these evidences by pointing that most of these works have been performed in an inactive state of membrane.

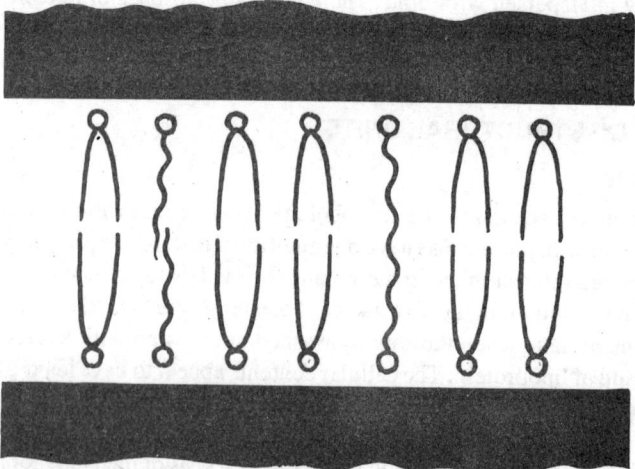

Fig. 5.2   Lamellar model of membrane showing hydrocarbon chain centrally located. Polar head groups of lipids are directed outward interacting with charged surface protein.

The corpuscular or sub-unit model defines a sheet of lipoprotein units, one unit thick, held together by hydrophobic interactions. The protein is found partially in the α-helical forms and partially as random coils. The hydrophilic peptide with segments are located at the hydrated membrane surface and interact with other polar peptide chains and polar head groups of lipids. Thus the external surfaces of the sub-units are polar in character containing polar peptide chains and polar lipids. Within the sub-units are hydrophobic peptide chains packed amid the hydrocarbon lipid chains. In Green's model the subunits are held together by protein-protein hydrophobic interaction. These are visualised as cubes with two opposite faces occupied by lipid, allowing only the four protein faces to interact to give a sheet-like membrane.

Of the two models there is much evidence favouring the sub-unit model. In general, it allows for easy accommodation of the permeability and enzymic functions found in the membrane. It shows no impassable

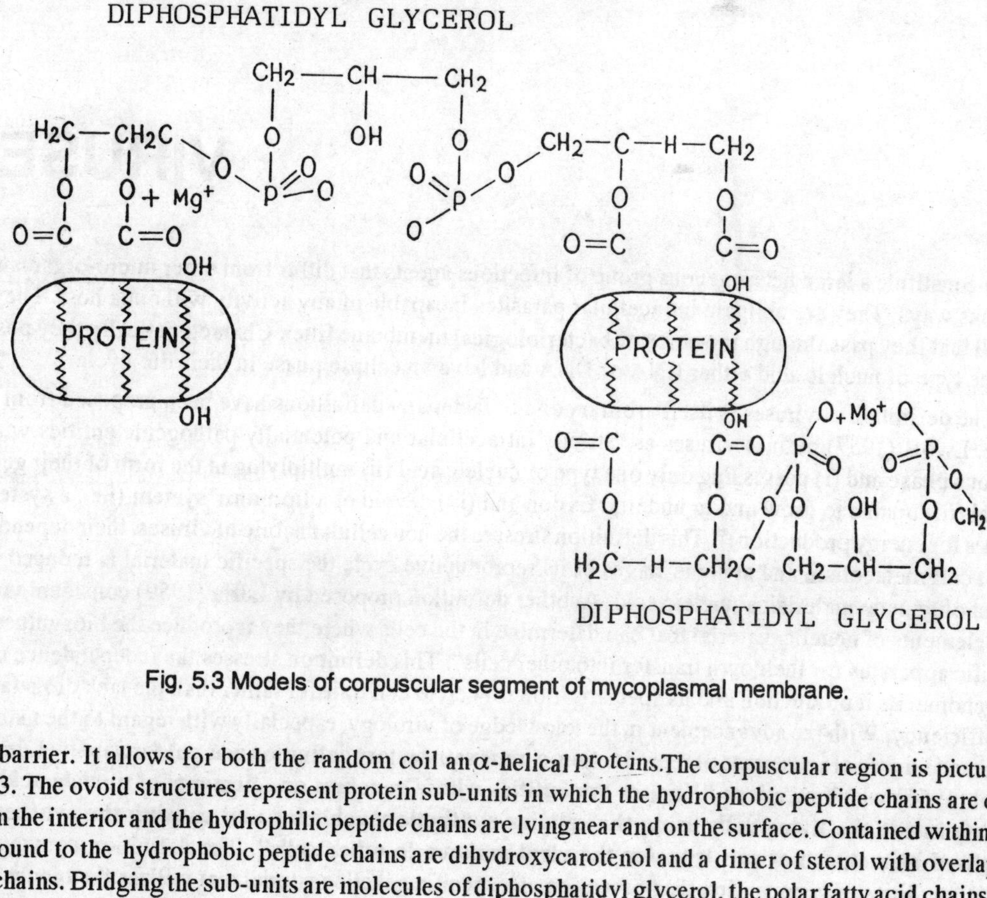

DIPHOSPHATIDYL GLYCEROL

Fig. 5.3 Models of corpuscular segment of mycoplasmal membrane.

lipid barrier. It allows for both the random coil an α-helical proteins.The corpuscular region is pictured in fig.5.3. The ovoid structures represent protein sub-units in which the hydrophobic peptide chains are coiled within the interior and the hydrophilic peptide chains are lying near and on the surface. Contained within them and bound to the hydrophobic peptide chains are dihydroxycarotenol and a dimer of sterol with overlapping side-chains. Bridging the sub-units are molecules of diphosphatidyl glycerol, the polar fatty acid chains. This model is an over simplification of what the membrane may be in reality and is accepted as a workable model for detailed study of mycoplasmas.

SUGGESTED READINGS

Gibbons, N.E. (ed.) (1963). Recent progress in Microbiology. Univ. of Toronto Press, Toronto.

Hayflick, L. (1969). The Mycoplasmatales and the L-phase of bacteria. Appleton, New York.

Raychaudhuri, S.P. (ed.) (1979). Mycoplasma disease of trees. Associated Publishing Co. New Delhi.

Roberts, R.B. (1968). In 'Microbial protoplasts, spheroplasts and L-forms' (L.B. Guze ed.). Williams and Wilkins, Baltimore, Maryland.

Smith, F. (1971). The biology of Mycoplasma. Academic Press, New York.

Whitcomb, R.F. and J.G. Tully (eds). (1978).The Mycoplasmas. Vol.III. Academic Press, New York.

6

VIRUSES

Viruses constitute a large heterogenous group of infectious agents that differ from other micro-organisms in numerous ways. They are obligate intracellular parasites incapable of any activity without a host. They are so small that they pass through the pores of bacteriological membrane filter. Characteristically they possess only one type of nucleic acid either RNA or DNA and have an eclipse phase in their life-cycle.

The definition of viruses is itself arbitrary and in fact many definitions have been proposed from time to time. Lwoff (1957) defines viruses as "strictly intracellular and potentially pathogenic entities with an infectious phase and (i) possessing only one type of nucleic acid (ii) multiplying in the form of their genetic material (iii) unable to grow and to undergo fission and (iv) devoid of a **lipmann system** (i.e., a system of enzymes for energy production)". This definition stresses the non cellular nature of viruses, their dependence on host cell metabolism and at some stages of its reproductive cycle the specific material is reduced to an element of genetic material, a nucleic acid. Another definition proposed by Luria (1959) considers viruses to be "elements of genetic material that can determine in the cells where they reproduce the biosynthesis of a specific apparatus for their own transfer into other cells". This definition stresses the independence of the viral genome, its reproduction and its specialisation for cell to cell transfer rather than the lack of metabolic self sufficiency. With the advancement in the knowledge of virology, especially with regard to the nature of their genetic material and events of reproduction, a more satisfactory definition was put forward by Luria and Darnell (1967) which mentions **"Viruses are entities whose genomes are elements of nucleic acid that replicate inside the living cells using the cellular synthetic machinery and causing the synthesis of specialised elements that can transfer the viral genome to other cells".** The definition conveys two essential qualities of viruses; first the possession of a specific genetic material that utilizes the biochemical machinery of the host cell and second the possession of an extracellular infective phase represented by specialised objects or **virions** which are produced under the genetic control of the virus and serve as vehicles for introducing the viral genome into other cells. Viruses thus qualify for **intracellular parasitism,** strictly speaking, parasitism at the genetic level, that regulate the synthesis of extracellular infective phase, the virion.

Before we discuss the nature of viruses in detail it would be appropriate to consider two basic questions about viruses. Are viuses living ? Are they organisms ? A virion in purified form possesses uniform size, shape and chemical compositon and can even crystallize (a non-living character) and at the same time has ability to reproduce (a living character). When we consider the words 'organisms' and 'living' these have unambiguous meaning mainly applied to the objects for which they are actually coined. An earthworm is an organism. A jumping cat is a living organism. But what actually makes an earthworm or a cat an organism? Lwoff (1957) has defined organism as "an independent unit of integrated and interdependent structures & functions". An earthworm is such a unit, but the cellular organelles - mitochondria, chromosome, chloroplasts are not organisms because they lack independence. According to this definition a virus would not be an organism because it is not independent; it depends on a living cell for the expression and replication of genetic material. We may consider another definition of organism that stresses individuality, historical continuity and evolutionary rather than functional independence. "An organism is the unit element of a continuous lineage with an individual evolutionary history (Luria *et al.*, 1977)". A virus attains a relatively independent evolutionary history by its peculiar adaptation to transfer from host to host. It can widen its range of variants that are not confined within any one lineage of host organism. A virus has definitely more independence then any cellular organism. Thus, evolutionarily speaking, it is more an organism than a mere gene-complex, though functionally it is much less independent than any such cell.

The living and non living nature of viruses can also be viewed in the light of above explanation. A life is a property, a manifestation or a state of cells and organisms exhibiting independent unit of structure and function. "A material is living if, after isolation, it retains a specific configuration that can be reintegrated into cycle of genetic matter (Luria, 1953)". Thus the living things possess an independent, specific, self-replicating pattern of organisation. Accordingly a protein is non living since its specific structures, its amino acids, can not serve as template to be copied by any cell; but the specific base sequence of the nucleic acid of a gene can be copied and hence is living. Thus a virus is also alive. Transferability of a viral genome is an essential part of viral existence. And this transferability is a reasonable basis for considering virus 'more living' than any other fragments of genetic material, in the same way as it is a reasonable basis for considering them 'more organism' than any cellular organelles.

NATURE OF VIRUSES

Size

Viruses are distinguished by their **small size**. The largest animal virus e.g., smallpox virus measures about 300 nm in diameter whereas the smallest one like "foot and mouth disease virus" of cattle reaches only 10 nm in size. On an average the viruses are 100 nm or even lesser (Fig. 6.1). The structure and organisation of viruses are studied with the help of a number of sophisticated equipments using modern techniques (See Box-.2)

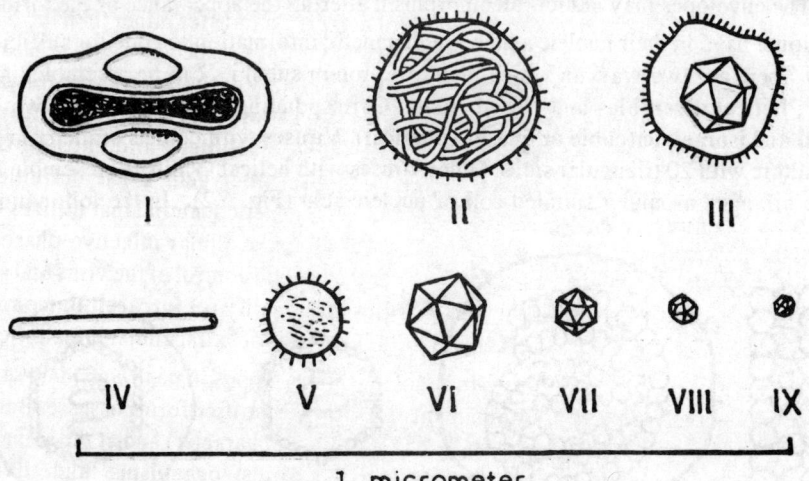

Fig. 6.1 Morphological forms and size of various virus particles.

i. pox virus ii. mumps virus iii. herpes virus iv. tobacco mosaic virus v. influenza virus vi polyhedral insect virus vii. adenovirus viii polyoma virus ix. poliomyelitis virus.

Box 2

SOME SOPHISTICATED TECHNIQUES FOR VIRUS STUDY

Electron microscopy - Electron microscopy with a resolving power of 0.4 to 1 nm. (4 to 10 A°) reveals the detailed structure of the particles of viruses. It produces photographs of materials dried from the frozen state or prepared by the critical point method.

Shadow casting technique - An opaque material gold, pallidium, uranium is vaporised at a known angle to the plane of the specimen. The shadow of the objects reveals its height and shape.

Replica technique - A mold or replica of the object is made by pouring a solution of plastic mateiral over the object. When the plastic is dry it is stripped off and photographed.

Negative staining - Drying a suspension of virions in a solution of phosphotungstate or uranyl acetate which on the electron opaque background reveals the finest structural details.

X - ray diffraction - Purified preparation of virions that form either true crystal (e.g., poliovirus) or paracrystal i.e., oriented gel of rod-shaped particle (e.g., TMV) land themselves to structural analysis by X - ray diffraction. It gives information on the shape and internal structure of virus and uncovers the structural units that make up the particles.

Organisation of Virion

Most viruses have relatively simple structure. The basic stucture of all virions is the **capsid** which encloses only one of the two types of nucleic acids. Some infectious agents, the **viroids**, which cause variety of plant diseases (e.g., potato spindle tuber) have no capsid and consist only a low molecular weight RNA. Capsids are made up of protein sub-units called **caposmere** which are assembled according to simple geometrical principles based on elementary physical consideration. The capsid and the nucleic acid core form the **nucleocapsid**. The large and more complex viruses often have an additional covering, the **envelope**, surrounding the capsids. The envelope is phospholipid or lipoprotein bilayer derived from host cell membrane that have been modified by the addition of virus protein. In some envelopes the proteins appear as projection called **spikes**. The envelopes may exhibit pleomorphism altering the appearance of the virion.

Viruses donot have in their nucleic acid enough genetic information to code for sufficient proteins to make a capsid. There are two ways in which identical protein subunits can be assembled to build stable, regular capsid ; **helical assembles** and **closed shells**. Correspondingly there are only two basic types of capsids' **helical** and **isometric** (cubic or quashi-spherical). Viruses with cubic symmetry are **icosahedral**, sphere like structure with 20 triangular sides. Other viruses with helical symmetry resemble long rods with the capsomere arranged around a spiraled coil of nucleic acid (Fig. 6.2). In the following sections four

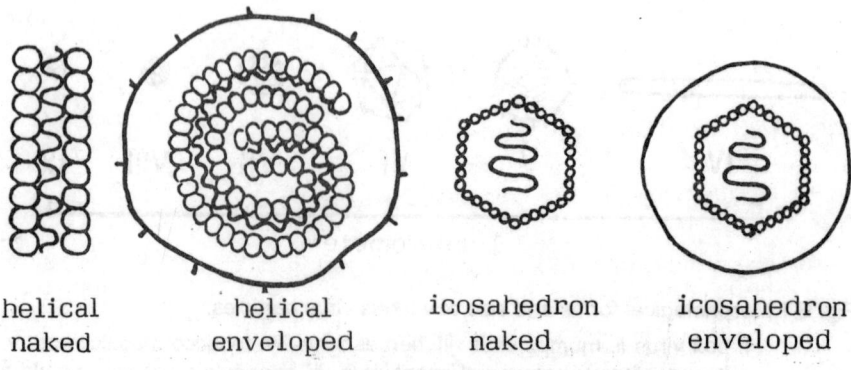

| helical
naked | helical
enveloped | icosahedron
naked | icosahedron
enveloped |

Fig. 6.2 Structural forms of virus particles.

representative viruses of each group having different morphologial organisations i.e. virons with helical symmetry; naked (TMV) and enveloped (Influenza virus); and virions with cubical symmetry; naked (Poliomyelites) and enveloped (Herpes virus) will be discussed. Table 6.1 presents a general survey of viruses according to their structural properties.

Table 6.1
MORPHOLOGICAL CATEGORIES OF VIRUSES

Naked Capsids	Enveloped Capsids
Helical symmetry	
RNA - Tobacco Mosaic Virus and many other plant viruses.	RNA - Myxovirus . Influenza virus, Parainfluenza virus, mumps and measels virus
DNA - Coliphage d.	
Polyhedral Symmetry (icosahedron)	
RNA - Picornavirus, Foot and mouth disease virus, Polio Reco- virus	RNA - Retrovirus : (oncovirus) sarcoma, leukaemia, carcinoma
Cyclic	
DNA - Papovavirus : polyoma	(onco-virus) papilloma,
Simian virus (SV40)	DNA - Herpes simplex virus
Single stranded cyclic	
DNA - Coliphage ϕ X 174 , M113	Varicella (Chicken pox)
	Epstein - Barr (EB-virus)
Composite Viruse (icosahedral head + helical tail)	
DNA bacteriophage (T_2, T_4, T_6)	
Complex virions DNA	Poxvirus (Variola)
Cow pox (Vaccinia)	

Helical naked

The virions of many plant viruses and several phages have naked helical capsids. Electron microscopy of Tobacco Mosaic Virus (TMV), the best known virus of this group, shows rods 15 to 17 nm thick. There may be some variations in length but only those rods that are 3000 A° long are infectious. The virions have a hexagonal contour with a central hole 40 A° in diameter.

X-ray diffraction of TMV shows an axial repeat of 69 A°, reflecting a left handed helical assembly of 49 protein subunits for every three turns (Fig. 6.3). A virion contains 2100 (± 3%) identical protein subunits; each protein molecule is 17,400 in molecular weight. The protein subunits are tapered on the outside forming an external groove. The amino acid sequence of the protein molecules is fully known. The virion has a molecular weight $39 \times 10^6 \pm (3\%)$ and contains a single RNA molecule weighing 2.06×10^6 daltons. The TMV protein aggregate in the form of flat disk of 17 subunits each. The flat disk is a more open structure than the helical assembly. It appears that disks serve as primer to initiate the assembly of the virion. Upon assembly the disks undergo a transition from the flat shape to a lock-washer shape corresponding to about one turn of the TMV helix.

The TMV capsids are rather rigid rods. The capsid of other plant viruses such as sugar beet yellows or potato virus X are helical but flexible rods.

0   10nm

Fig. 6.3   Diagram showing arrangement of rod-shaped TMV virion. Note the helical arrangement of large shoe-shaped protein structure and inner RNA beaded structure.

Helical enveloped

The particles of influenza virus have a diameter 110 nm. The nucleocapsid is helically arranged as in TMV, though it is not rod shaped but twisted or rolled up. The nucleocapsid is surrounded by an envelope which is part of the membrane of the host cell from which the virion originated. The envelope has spikes on its outer surface; these serve to absorb the virion to the host cell and contain protein and the enzyme neuraminidase. Multiplication of the virus takes place inside the host cell whilst the liberation of the virions occurs by a kind of budding.

Polyhedral viruses without envelope

Many viruses that appear spherical are actually polyhedral. The preferred polyhedral form is the **icosahedron** (20 faces), a body of 20 equilateral triangles and 12 vertices. The capsid of such viruses consists of two types of capsomere; in the vertices there are five cornered pentons consisting of five proteins (**protomers**), the planes and edges are occupied by six cornered hexons consisting of six protomers. The assembly of the capsid from the capsomeres follows the laws of crystallography; the smallest icosahedral capsid would consist of 12 pentons, the next larger one would have 12 pentons and 20 hexons. There are viruses that have 252 and even 812 capsomeres. The poliomyelitis virus, the virus of foot and mouth diseases, adenovirus and the SV40 tumor virus are the common examples of this group.

Polyhedral viruses with envelope

The architectural form of an icosahedron surrounded by an envelope is found as the causative agents of chicken-pox (Varicella), shingles (Herpes Zoster) and vesicular stomatitis.

The icosahedral capsid of herpes virus consists of 162 capsomeres. The envelope is derived from the inner nuclear membrane of the host cell. The capsids are covered by the inner nuclear membrane of the host, liberated by budding and guided to the outer surface by the endoplasmic reticulum.

Pox virus

The pox viruses are the largest among the animal viruses. Their construction is quite different from the four types mentioned above. They contain DNA, proteins and several lipids. They are, therefore, called complex virions. The virus particles of small pox (Variola) and of cow pox (Vaccinia) have the shape of a rounded off square. They consist of an inner body which contains the double stranded DNA, a protein-containiug double layer elliptical protein bodies and a covering membrane; all of this is covered by closely applied threads. The virus particles are very resistant to dehydration and hence very infectious. Both viruses, vaccinia and variola, have common antigens.

Nucleic acid

The amount of nucleic acid in viruses varies considerably e.g., the virion of influenza virus contain 1% RNA, tobacco mosaic virus 5% RNA, small pox virus 5% DNA and some of the bacteriophages as much as 50% DNA. Accordingly the molecular weight of the genome shows great diversity among viruses. Some viurses like small pox may have 300-400 genes whilst others like SM-2 may only have three genes. Viruses of polio, influenza, measels and tobacco mosaic have their sole genetic as RNA. The physical configuration of the genome also varies with different viruses. Thus four possible combinations may exist (1) double stranded DNA (2) single stranded DNA (3) double stranded RNA (4) single stranded RNA. In the case of DNA viruses the molecule may be linear or a covalently closed circle. In some viruses, e.g., poliovirus, the genome exists as a single molecule in the virion whilst in others e.g., influenza virus, the genome is segmented i.e., the virion contanis segments of RNA.

Viral protein

The viral protein forms the largest fraction (50 to 90%) of the virion. The main function of protein coat or capsid is to protect the delicate core of nucleic acid. In addition the capsid proteins may also provide specific receptors for attaching the virus to the host cell. Besides structural protein some viruses also contain enzymes e.g., some phages contain the enzyme **lysozyme** which facilitates the entry of the phages nucleic acid through the bacterial (host) cell wall. In other viruses, the enzymes are usually nucleic acid polymerases e.g., influenza virus has an RNA dependent RNA **polymerase**, small pox virus a DNA dependent DNA polymerase. Tumor viruses have an RNA dependent DNA polymerase (**a reverse transcriptase**).

Host specificity

Viruses are obligate intracellular parasites. For replication they rely entirely on the metabolic processes of the host cell, particularly those associated with nucleic acid replication and translation of m-RNA. Plants, animals and micro-organisms are the natural hosts of viruses.

Plant viruses.

Plant viruses get entry to their hosts through lesions; they do not actively penetrate into plant tissues. In nature these viruses are spread by vectors or by direct contact. The plant parasite *Cuscuta* can also form a conducting system for virus infection over plants through their haustoria. Transmission of plant virus is detailed elsewhere. Most plant viruses have RNA as their genetic material.

Animal viruses

In humans and other animals, viruses cause many diseases (see p 141). Animal viruses are also transmitted by direct contact or via insects. They apparently gain entry into the host cell by phagocytosis or pinocytosis. The genetic material of animal viruses can be either DNA or RNA. Whilst DNA is usually present as a double stranded helix, RNA is found as a single stranded polynucleotide chain.

Bacterial viruses

Viruses that use bacterial cells as hosts are called bacteriophages. The presence of bacteriophages is recognised by the appearance of **plaques** or lytic holes in a continuous bacterial lawn. Phage nucleic acid occurs either as double or single stranded DNA or as single stranded RNA. The phages of *Escherichia coli*

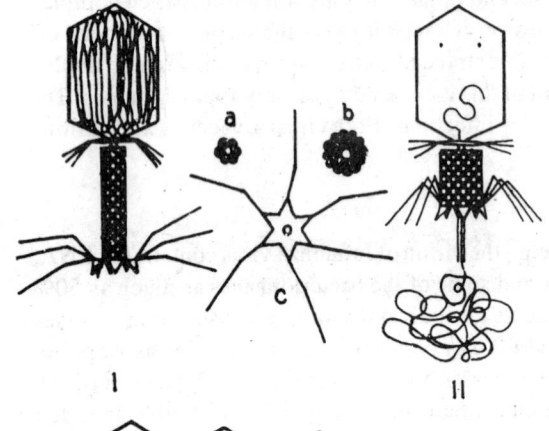

Fig. 6.4 Model of a T_2 phage.

i. Phage with stretched sheath before adsorption

ii. Phage with contracted sheath after adsorption

a) Transverse section through stretched tail; 6 sheathed protein unit in one plane

b) Transverse section through the contracted tail; 12 - sheathed protein unit in one plane

c) view of the basal plate of phage, ready for adsorption.

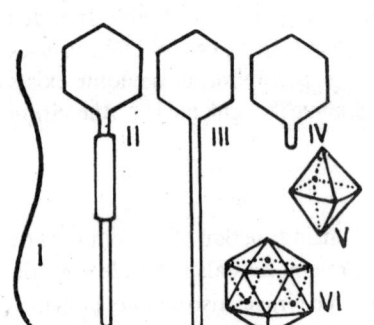

Fig. 6.5   Various shapes of bacteriophages (i-iv) and geometrical shapes of phage heads (v-vi).

i. Thread-like form of coliphage fd.

ii. Hexagonal head with contractile sheaths (eg; T_2, T_4, T_6)

iii. Head with long flexible non-contractile tail (eg; T_1, T_5)

iv. Head with short tail (eg; T_3, T_7)

v. Octahedron vi. Icosahedron.

The shapes and forms of bacteriophages have been elucidated mainly for T-series of *E. coli* phages. The coliphage T_2 consists of a polyhedral head, about 100 nm in length and a tail (Fig. 6.4). It is called a composite virus. The head comprises capsomere and contains DNA; protein and DNA each comprises about 50 % of the head. The tail has a rather complicated structure consisting of three parts. A hollow **stylus** is surrounded by a contractile sheath which bears on its distal end a base plate covered with claw like tail fibre and host specific adsorption spikes. According to their form and structure the T-series bacteriophages of *E. coli* have been numbered as T_1, T_2...T_7 etc. The morphology of these bacteriophages ar presented in fig. 6.5.

Algal viruses

Many of the blue-green algae are attacked by viruses that are known as cyanophages. They were first discovered in 1963 by Safferman and Morris. These groups are usually designated by the initials of the generic names of the corresponding hosts to which arabic numerals are added for designating the serological sub-groups. In morphology they resemble the bacteriophage. Grouping of cyanophages is done on the basis of their host specificity, morphological and serological properties (Table-6.2).

Table 6.2

MAIN GROUPS OF CYANOPHAGES MORPHOLOGY

Cyanophage group	Specific Host	Head	Tail
LPP-1	*Lyngbya, Phormidium Plectonema*	Hexagonal 60 ± 2 nm in diam	Short non contractile
SM-1	*Synechococcus elongatus* and *Microcystis aeruginosa*	Icosahedral 67 ± 1.8 nm	Short, collar with thin appendages
N-1	*Nostoc muscorum*	Hexagonal 55 nm	Long, contractile
AS-1	*Anacystis nidulans and Synechococcus cedroum*	Hexagonal	Long, contractile

Fungal viruses

Viruses attacking fungi are now well recognised. They have been reported in species representing each of the major taxonomic classes of fungi (Table 6.3). The most extensively studied system are the mycovirus of *Penicillium crysogenum*. They are of icosahedral symmetry that contain double stranded RNA. Detailed studies of fungal viruses are, however, lacking due to difficulties in measuring viral titer as well as cellular complexity of the eukaryotic fungal host.

Table 6.3

DISTRIBUTION OF FUNGAL VIRUSES IN SOME REPRESENTATIVE SPECIES OF FUNGI

Taxonomic Group	Fungal species
PHYCOMYCETES	*Mucor* spp., *Rhizopus* spp., *Plasmodiophora brassicae*
ASCOMYCETES	*Neurospora crassa, Peziza ostracoderma,* Saccharomyces carlsbergensis, S. cerevisae,
BASIDIOMYCETES	*Agaricus bisporus, Boletus* sp. Coprinus lagopus, Puccinia graminis, Ustilago maydis
DEUTEROMYCETES	*Alternaria tenuis, Arthrobotrys* spp., *Aspergillus flavus, A. glaucus, A. niger, Botrytis* spp., *Candida utilis, Fusarium moniliforme,* Helminthosporium maydis, Penicillium crysogenum, P. stoloniferum, Verticillium spp.

REPLICATION OF VIRUSES

The replication of viruses in their hosts is a very complicated process. This differs from the multiplication of other micro-organisms which exhibit binary fission with or without mitotic division of their genetic components. The mode of entry of viruses into their respective hosts varies widely. The plant viruses and the bacteriophages have to overcome the obstacles of the cell wall in order to penetrate the host cells. In case of plant viruses the cell wall is breached either by trauma e.g., grafting or as a result of direct injection by virus-carrying insect vectors. In animals the entry of viruses is either through direct contact or through droplet nuclei and in the case of phages only the virus nucleic acid penetrates the bacterial cell wall by an elegant injection device associated with the phage tail. The studies on replication process of viruses have been extensively carried out by the biochemical, genetic and morphological investigations of the bacteriophages of the T-series (T_2, T_4, T_6). Tissue culture studies have also revealed the nature of replication sequence occurring in the virus infected host cell. This includes :

(i)   adsorption of the virion to the cell plasma membrane

(ii) penetration of the virions into the cell either by micropinocytosis or by fusion of the viral envelopes with the cell membrane.

(iii) eclipse of the virion due to enzymically controlled decapsidation process.

(iv) biosynthesis of new virus components.

(v)   assembly of immature virus in the cell.

(vi) maturation and release of progeny virions from the infected cells either by budding off (e.g., influenza virus) or by cell lysis (e.g., poliovirus).

The host cell must be capable of supporting this sequence of steps in viral replication.

Based on the relationships with the bacterial cell, phages are of two types : (i) **virulent** or lytic phages that cause lysis of the host cells and (ii) **temperate** phages that may lyse the host but sometimes their genetic material becomes associated with that of the host cell as **prophage** and is transferred to daughter cell on division. The bacterial cultures which carry the temperate phages are termed **lysogenic.** Excision of the phage DNA can occur spontaneously or by treatment with UV light or mutagenic agents which, thereafter, undergoes for lytic multiplication.

The replication process of bacterial virus can be discussed under the followng lines :

Adsorption of Free Phage

Not every phage is adsorbed by every bacterium. The specificity of adsorption depends on the specific host phage relationship. For successful adsorption the phage particles must first collide with the cells and the complementary bonds must establish between the two surfaces. Under optimum condition the process is very rapid and is accomplished within a few minutes. Frequency of collision between phages and bacteria depends on the concentration of the virus particles as well as the mutual attraction or repulsion brought about by electrostatic forces. Available evidences indicate that both phage and host cell carry a net negative charge which allows the phage to approach close enough the cell. The formation of bonds after collision is dependent on the chemical structure of the two surfaces e.g., with *E. coli* T_1, amino groups on the viral surface bond with receptors containing acidic groups on the cell surface. The receptor site of the bacterial cell virtually determines the resistance or susceptibility of the bacterium to the phage. Such receptors may be situated in different layers of the wall e.g., receptors for phage T_2 and T_6 exist in the lipoprotein layer whereas those for T_3 and T_4 are present in lipopolysaccharide layer. Accodingly the phage surface must also be modified before adsorption. This may be attained by attachement of positively charged cations (Mg^{++} of L-tryptophane).Some phages (T even phage) are equipped with fibrils which get adsorbed to the receptors on the bacterial cell leaving the virus particles in a characteristic position at right angles to the cell wall and with the head sticking out. Some other phages attach specifically to flagella of the bacterium. Sometimes mutation can change the nature of chemical groups on the receptors thereby preventing phages adsorption and rendering the cells phage resistant. If phages are added in excess, multiple adsorption upto 200-300 phages per host cell can occur.

Penetration of Phage Nucleic Acid

After a virion attaches to a susceptible host there occurs a series of reactions that leads to the release of the genetic materials within the cell. The release of nucleic acid may be at the cell surface or some phages through these organelles. Sometimes enveloped virions may fuse with the cell membrane and release into the cellular cytoplasm the inner capsid which in turn releases the nucleic acid genome. In phage T_2 the injection of DNA involves the anchoring of the baseplate and contraction of the sheath which effects penetration of the hollow stylus into the bacterial cell. Experiments with ^{32}P-labelled nucleic acid and ^{35}S-labelled proteins have shown that only the nucleic acid penetrates inside the host cell while the protein coat is left outside. The T-even phages are known to hydrolyze the mucopeptide of the bacterial cell wall by means of a phage lysozyme which is attached to the tip of their tails. Sheath contains the source of energy for inserting the tube through the wall and does not depend on an energy supply from the bacterium. However, other phages which are not equipped with contractile sheaths are still able to effect penetration though at much slower rate and are dependent on the energy supply by the host.

Intracellular Development of Phages

Soon after the injection of phage nucleic acid into the bacterial cell, the presence of phage can not be demonstrated for a short while in the ruptured bacteria. This short period before phage can once more be demonstrated is called the **eclipse phase.** Once the viral genome is released from the virion it becomes available for replication and transcription serving as template for the biosynthesis of specific products. The synthesis of bacterial DNA stops abrruptly. Synthesis of bacterial RNA and protein also ceases but the total protein content continues to increase and DNA synthesis is resumed at a higher rate. At first, phage DNA is synthesised at the expense of degraded bacterial DNA. The enzymes needed for DNA synthesis are made shortly after injection; these are the so called **early protein.** The **late proteins** include the coat proteins for the phage head and tail as well as the phage lysozyme; these are only made during the second half of the late period. Genetic recombination in which phage particles undergo random exchange of genetical material can occur at this stage.

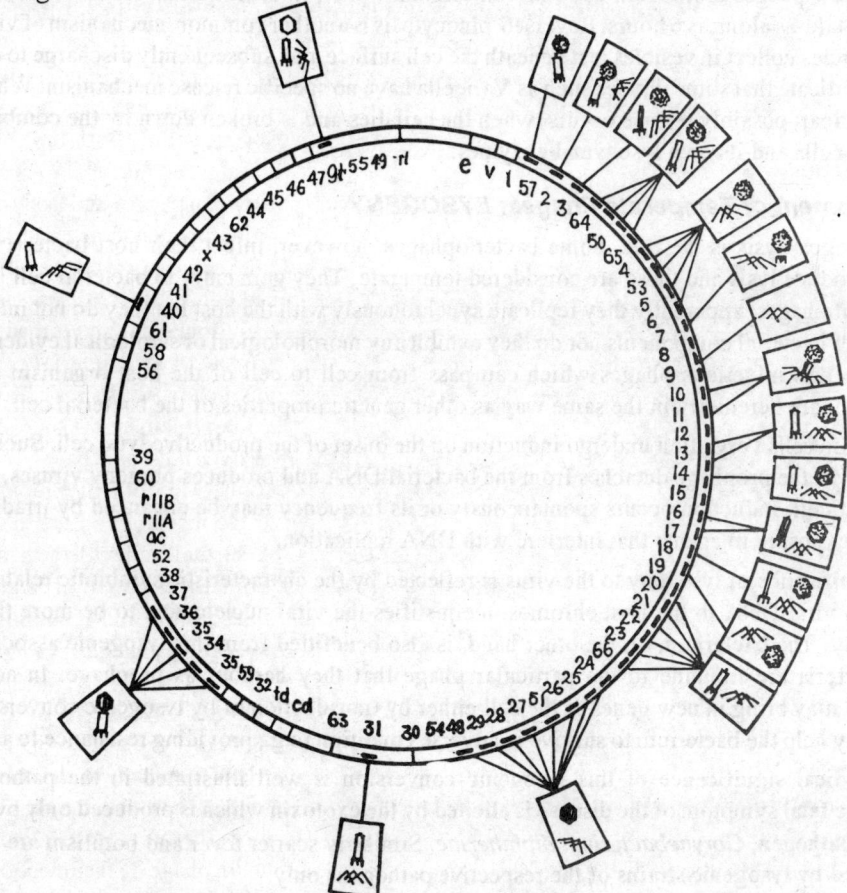

Fig. 6.6   A genetic map of the T₄ virus. The squares show the morphological element whose production is governed by a particular gene.

Maturation and Release of New Phages

The fundamental step in virus replication is the production of new molecules of nucleic acid which are exact replica of the nucleic acid in the infecting particles. Replica of nucleic acid and synthesis of viral protein occur almost simultaneously. Maturation of phages involves assembly of these components which occur in several steps. Each step is controlled by genes in the phage nucleic acid. In constructing coliphage of the T-even, three different pathways are involved in the formation of head, tail and tail fibres. Genetic map of the T4 virus (Fig. 6.6) shows the location of more than 75 genes of which 40 to 50 'architectural' genes control the synthesis and assembly of new virus. The pathway by which the head, tail and tail fibres are synthesised and assembled are outlined in Fig. 6.7. Apparently viral DNA molecules are first condensed into large particles surrounded by a membrane and resembling the heads; phage heads are completed by further condensation of capsid monomers around the DNA condensate. The completed heads are filled with the amount of DNA that is specific for each phage type and the heads are then closed. The hollow tube and the base plate appear to be assembled next, followed by the contractile sheath. Finally the fibrils are added.

In most of the phages, synthesis of phage components and assembly of mature phages continue unitl the bacterial cell bursts. The bacterial cell wall is softened by the phage lysozyme and the cell bursts releasing newly formed infective phages into the surrounding medium. The whole cycle of replication from adsorption to release of new phages takes from 20 to 30 minutes. However, in some animal viruses the the complete sequence may take as along as 6 hours. Reversed pinocytosis is another common mechanism of viral release. The virus particles collect in vesicles just beneath the cell surface and subsequently discharge to exterior. *In vitro* studies indicate that some viruses such as Varicella have no specific release mechanism. What happens *in vivo* is not clear; possibly release occurs when the cell dies and is broken down by the combined action of phagocytic cells and its own lysozymal enzymes.

The Development of Temperate Phages: LYSOGENY

In virulent phages, lysis is the rule. Some bacteriophages, however, infect their host bacteria but do not multiply or produce lysis and these are considered temperate. They gain entry to bacterial cell in the same way as virulent phages; apparently they replicate synchronously with the host but they do not interfere with the synthesis of bacterial components nor do they exhibit any morphological or serological evidence of their presence. The non infectious phages which can pass from cell to cell of the host organism are termed **prophage**; they are hereditary in the same way as other genetic properties of the bacterial cell.

Lysogenic cells very often undergo induction on the onset of the productive lytic cell. Such induction occurs whenever the prophage detaches from the bacterial DNA and produces progeny viruses, eventually lysing the host cell. Induction occurs spontaneously or its frequency may be enhanced by irradiation with ultraviolet or exposure to agents that interfere with DNA replication.

The significance of lysogeny to the virus is reflected by the characteristic symbiotic relatioship. The integration on viral DNA to the host chromosome justifies the viral nucleic acid to be more than a mere chemical entity. The bacterium, on the other hand, is also benefitted from this lysogenic association. The lysogenic bacteria are immune to the particular phage that they harbour as prophage. In addition the bacteriophage may bring in new genes to the cell either by transduction or by lysogenic conversion. These new genes may help the bacterium to survive in adverse condition (e.g., providing resistance to antibiotics).

The medical significance of this lysogenic conversion is well illustrated in the pathogenesis of diphtheria. The fatal symptom of the disease is elicited by the exotoxin which is produced only by lysogenic strain of this pathogen, *Corynebacterium diphtheriae*. Similarly scarlet fever and botulism are the serious diseases caused by lysogenic strains of the respective pathogens only.

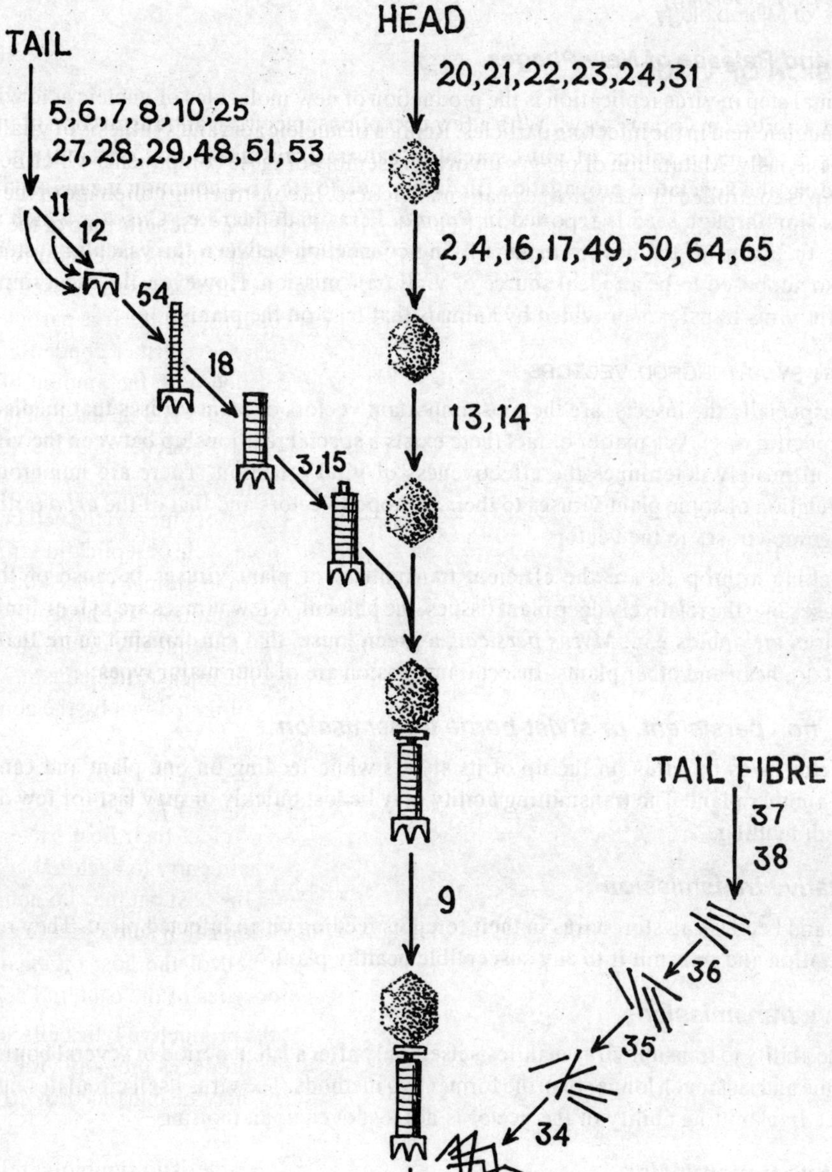

TAIL

5, 6, 7, 8, 10, 25
27, 28, 29, 48, 51, 53

11

12

54

18

3, 15

HEAD

20, 21, 22, 23, 24, 31

2, 4, 16, 17, 49, 50, 64, 65

13, 14

9

TAIL FIBRE

37

38

36

35

34

LABILE FACTOR

Fig. 6.7 Morphogenetic pathway illustrating the branches which combine to form complete virus particles.

TRANSMISSION OF VIRUS

Viruses are transmitted in various ways. With a few exceptions, mechanical transmission of viruses into the healthy plants is the major source of virus spread in nature. Transmission by contact (e.g., TMV) or by grafting, budding and vegetative propagation (in dahlia, potato etc.) is a common means of virus infection. Viral transmission through seed is reported in *Pitunia*. Parasite doddar i.e., *Cuscuta* which sends feeding haustoria into the stems of the host to make a living connection between the vascular system of host and parasite is also supposed to be an ideal source of viral transmission. However, the most important natural means of plant virus transfer is provided by animals that feed on the plants.

TRANSMISSION BY ARTHROPOD VECTORS '

Arthropods, especially the insects, are the most important vectors of plant viruses that mediate the transfer of virus in a specific way. As a matter of fact there exists a special relationship between the virus vector and plants which ultimately determines the effectiveness of viral infection. There are numerous similarities between the relation of some plant viruses to their arthropod vectors and that of the *arbo* (arthropod borne) animal pathogenic viruses to the vectors.

The sucking arthropods are the efficient transmitters of plant viruses because of their ability to introduce viruses into the relatively deep plant tissues, the phloem. A few viruses are xylem-limited. The most common vectors are aphids e.g., *Myzus persicae*, a green louse, that can transmit more than 50 different viruses of potato, bean and other plants. Insect transmission are of four major types:

1. *External, non persistent, or stylet-borne transmission*

The vector takes up virus on the tip of its stylets while feeding on one plant and can immediately transmit it to another plant. The transmitting ability may be lost quickly or may last for few days but is not carried through molting.

2. *Regurgitative transmission*

Many aphids and beetles can store virus in their foreguts feeding on an infected plant. They retain the virus for longer duration and transmit it to any susceptible healthy plant.

3. *Circulative transmission*

In this case the ability to transmit virus manifests itself only after a latent period of several hours or days from the feeding time and last much longer than the former two methods. The virus itself circulates inside the insect tissues and the transmitting ability of the vector is not lost even upon molting.

4. *Propagative transmission*

This mode of transmission is almost the rule with a large class of virus vectors. The virus, in such cases, multiplies in the insect tissues before reaching the mouth parts and within the latent period it undergoes multiplication. The virus eventually reaches the salivary glands and is inoculated into the plants. Leaf hopper which includes the vector of wound tumor virus, is the best studied example where the virus first multiplies in the filter chamber of the intestine and then after about two weeks appear in the hemolymph and in various organs.

Transmission of particular plant virus through insects is a genetically controlled phenomenon. However, the vector that serves as vehicle sometimes becomes victims of the virus infection. This has also raised certain intriguing problems -whether these are plant viruses or animal viruses ? Perhaps these have the ability of adopting to a 'double life' and may have some evolutionary significance.

TRANSMISSION BY NEMATODES

The role of nematode as vectors of several viruses including tobacco ring spot and tobacco rattle has been well ascertained. These worms transmit virus while feeding on roots and can harbour virus for months but do not transmit it to their progeny through the eggs. *Xiphinema, Longidours* and *Trichodorus* are the common nematode vectors. These are known as NEPO (Nematode Transmitted Polyhedral Virus e.g., Arabic Mosaic Virus) and NETU (Nematode Transmitted Tubular Particles e.g., Tobacco Rattle Virus).

TRANSMISSION BY FUNGI

Olpidium brassicae, a parasitic primitive phycomycetes is the best known example that transmits tobacco necrosis virus to the plants. The transmission is specific both with respect to virus strains and to vector strains and the virus is supposed to be carried on through the fungal zoospores. Vein diseases of lettuce and tobacco necrosis are reported to be transmitted by species of *Synchytrium.* Some plasmodiophorales e.g.. *Polymyxa graminis* is known to transmit wheat mosaic virus.

CLASSIFICATION OF VIRUS

Classification of viruses were initially based on the more easily observed features of virus diseases e.g., tissue tropism, host range symptomatology, pathology etc. International Committee on Taxonomy of Virus (ICTV) classified viruses on the basis of their host specificities e.g., viruses that infect plants only (tobacco mosaic virus, cyanophages, bacteriophages), arthropods (insect virus, cabbage looper), infecting arthropods and warm blooded vertebrates (eastern and western equine encephalitides), warm blooded vertebrates (measeles, influenza etc.), cold blooded vertebrates (Luck'e viruses of frog and other). Holmes (1948) grouped viruses under the order Virales which was divided into three suborders : **Phaginae** (infecting bacteria), **Phytophaginae** (infecting plants), **Zoophaginae** (infecting animals). These classifications, though convenient for purpose of discussion are arbitrary and incomplete and lack authenticity. In the year 1962 Lwoff, Horne and Taurnier proposed a system of classification which was acceptable to the International Association of Microbiological Society. The system, popularly known as LHT system, is based on the knowledge of physical and chemical properties of viruses that include the following essential features : -

(i)   Type of nucleic acid: DNA/RNA

(ii) Symmetry: helical, cuboidal or binal

(iii) Presence or absence of envelope around the nucleocapsid

(iv) Diameter of nucleocapsid, number of capsomere.

In the present system of classification all the viruses are grouped in the phylum Vira. This is subdivided into classes, orders, suborders and families. A schematic representation of the LHT system of classification is presented in Table-6.4. Properties that determine genera within families comprise :

(i)   Nucleic acid - base sequence, relative number of base, number of nucleotide.

(ii) Capsomere - structure, antigenic properties, molecular weight.

(iii) Capsid - number of capsomers, antigenic properties, reaction to heat, pH, other physical and chemical agent.

Table 6.4 LHT SYSTEM OF CLASSIFICATION OF VIRUSES

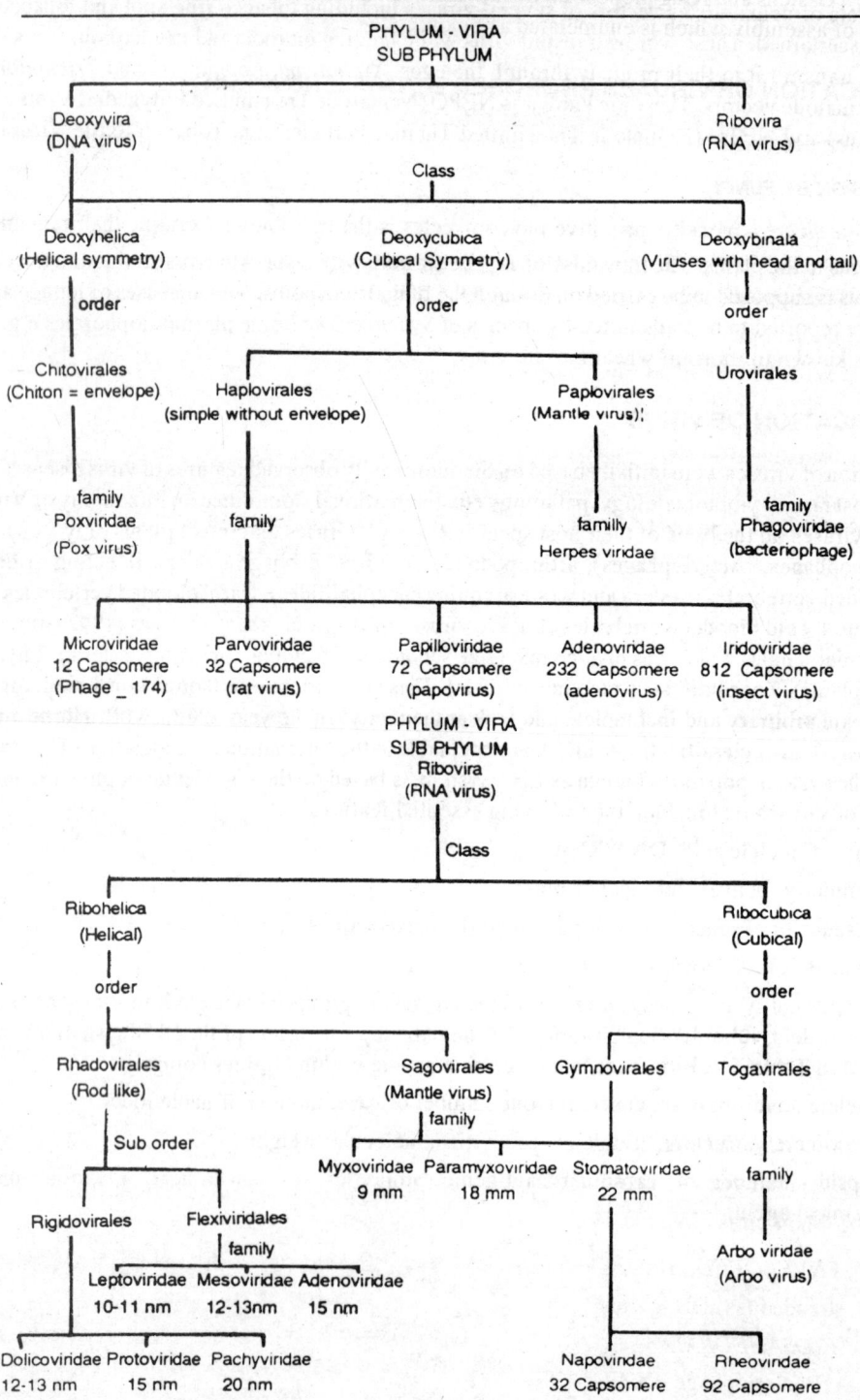

In 1975 Casjens and King proposed a fresh classification of viruses based on the nature of nucleic acid and the site of assembly which is enumerated as below :

CLASSIFICATION OF VIRUSES PROPOSED BY CASJENS & KING.

I. Single stranded (ss) RNA viruses

    A. *HELICAL*

        (a) Rigid rods (plants) - TMV, Barley stripe mosaic virus, Tobacco rattle virus.

        (b) Flexuous rods (plants) - Potato X and Y, Clover Yellow mosaic virus.

    B. *ICOSAHEDRAL*

        (a) Spherical plant viruses

          (i) With 180 identical capsomers (T=3)

            Cowpea Chlorotic Mosaic Virus (CCMV)

            Cucumber Mosaic Virus (CMV)

            Turnip Yellow Mosaic Virus (TYMV).

          (ii) With 60 subunits of 2 structural proteins (T=1)

            Cowpea Mosaic Virus

        (b) Bacteriophage - R_{17}, fr, f_2, MS_2, QB

        (c) Picornavirus (animal)

          (i) Human Entero Viruses - Polioviruses,

            Coxsachie and ECHO virus.

          (ii) Rodent Cardiovirus - Encephalomyocarditis Virus, Mengovirus

          (iii) Rhinovirus - Human respiratory infection

          (iv) Foot and Mouth disease virus

    C. *ENVELOPED*

        (i) Spherical - Togavirus - Yellow fever

        (ii) Bullet shaped - Rhabdovirus - rabies

        (iii) Spherical or filamentous - Paramyxovirus-measels.

        (iv) Spherical - Coronavirus, Arenavirus, Oncovirus.

II. Double stranded (ds) RNA virus

    A. *SEGMENTED GENOME*

        (a) Animal virus - Rheovirus, Blue tongue virus of sheep.

        (b) Plant virus - Wound Tumor virus, Maize Rough Dwarf Virus, Rice Dwarf Virus

    B. *ENVELOPED - BACTERIOPHAGE* λ 6

III. Single stranded (ss) DNA virus

    A. *ICOSAHEDRAL*

        (a) Bacteriophage - X 174, S 13

        (b) Parvovirus - Animals and insects

B. *HELICAL*

   (a) Bacteriophages - fd, f1, M13

IV. Double stranded (ds) DNA virus

   A. *ICOSAHEDRAL COMPLEX (tailed)*

      (a) E. coli phages - T4, P_2, P_4.

         T_3, T_7, T_5

      (b) *S. typhimurium* phage - p_{22}

      (c) *B. subtilis* phage - λ_{29} Blue - green algal virus, Bacteriophage - PM2

   B. *ENVELOPED*

   C. *NUCLEAR ASSEMBLY*

      (a) Papovavirus - Polyoma virus,

         SV 40, Human Wart Virus

      (b) Adenovirus - Respiratory disease in bird, mammals

      (c) Herpes virus - Cold sores, Shingles, Cervical sarcoma of uterus, Burkitt's lymphoma

   D. *CYTOPLASMIC ASSEMBLY*

      (a) Poxvirus - Variola Small Pox,

         Vaccinia immunity to small pox

NOMENCLATURE OF VIRUSES

Nomenclature of viruses has always been arbitrary. Although the virus families have been given latinised names, virologists have so far resisted the use of latinised binomial generic and specific names for individual virus. Perhaps it would take some more time to replace the vernacular names (e.g., tobacco mosaic virus) with scientifically accepted latinised names. Some workers favour the use of a binomial nomenclature based on the present vernacular name as well as an informative cryptogram. Recently a comprehensive nomenclature of plant viruses has been proposed by Henning P. Hansen (1981) on the basis of a simple latinised code-name system developed and published during the year 1956-1970. Each generic code name identifies the virus in question by direct information about its ways of transmission (including vector relations) and its type of particles. A brief summary of the symbols and the way they are combined to form code names is as follows:

  1. The initial indicates the normal regional site of infection.

     D (a) - (Only deep transmission) : to central region

     B (e) - (Mechanical transsision) : to leaf surface

     G (e) - (Soil transmission)   : to surface region

  2. The middle symbol indicates type of specific vector, if any.

     aca = mite; *in* = no specific vector

     ale = white flies; *nema* = nematode

     aphi = aphids; *plano* = fungal zoospores

     cica = leaf hoppers; *psylli* = psyllids

     het = plant bugs; *thy* = thrips

     Transmission by beetles (a = o) biting insects is considered non specific.

3. Ending symbols for type of particles, if any

virus : particles not defined

nodum : naked virus, free molecules only

(potato spindle tuber virus)

flexus : thread like flexuous

(potato X and Y virus)

lancea : thread like curved or slightly flexuous

(potato virus)

chorda : rigid rods, moderately long and thin

(tobacco mosaic virus)

pachus : rigid rods, relatively short and thick

(tobacco rattle virus)

media : bacilliform, moderately long and slim

(cocao mottle leaf virus)

physa : bacilliform, large sausage with a membrane often no end cut

(lettuce necrotic yellow virus)

globus : spherical, small, polyhedrons or globose

(turnip yellow mosaic virus)

sphaeria : spherical, large, polyhedrons or globose

(wound tumor virus)

sagum : roughly spherical with a membrane

(tomato spotted with virus)

myca : mycoplasma like agent

(aster yellow virus)

Example of combination of symbol to form a code name. Tobacco Mosaic Virus is coded as : *Minchorda nicotiana* — the virus exhibits mechanical transmission with no specific vector and is characteristically rigid rods, moderately long and thin. Similarly Beet Yellow Virus is coded as *Daphiflexus betae* which signifies deep transmission to central region by means of aphid as vector and is thread-like and flexuous.

SUGGESTED READINGS

Burke, D.C. and Russel (eds.) (1975). Control process in Virus multiplication. 25th Symp. Soc. gen. Microbiol. Cambridge University Press, Cambridge.

Crawford, L.V. and M.G.P. Stoker (eds.) (1968). The molecular biology of viruse. 18th Symp. Soc. gen. Microbiol. Cambridge University Press, Cambridge.

Casjens, S. and J. King (1975). Virus Assembly. **Ann. Rev. Biochem. 44, 363.**

Davis, B.D., R. Dulbecco, H.N. Eisen and H.S. Ginsberg (1980). Microbiology 3rd Edn.

Harper and Row, New York. Fenner, F. (1976). The classification and nomenclature of viruses. **J. gen. Virol. 31: 363.**

Hahon, N. (ed.) (1964). Selected papers on Virology. Prentice Hall, Inc. Englewood Cliffs, New Jersey.

Horne, R.W. (1974). Virus structure. Academic Press, London.

Luria, S.E., J.E. Darnell, Jr., D. Baltimore and A. Campbell (1978). General Virology, 3rd ed. John Wiley and Sons, New York.

Martin, S.J. (1978). The Biochemistry of Viruses. Cambridge University Press, Cambridge.

Smith, K.M. (1974). Plant Viruses. Champman and Hall, London.

7

GENETICS OF MICROBES

An organism resembles to its ancestors in most of its characters. The constancy of characters over the generation is called heredity. The distribution of characters in the progenies follows the **law of heredity** (first proposed by Johann Gregor Mendel) and the science dealing with the transfer of characters and variations among organisms is known as **Mendelian Genetics**. This classical genetics has undergone a radical change since the development of the concept of chemical structure of DNA, proposed by Watson and Crick in 1953 and during the last twenty five years it has flourished as a new branch of molecular biology on the pedastal of microbial genetics.

According to classical genetics the genes situated in the cell nucleus are arranged in linearly order. For a long time it was believed that genetic information was associated with the protein component of the nucleoplasm. However, the successful transfer of genetic information (transformation) by DNA revealed that this must be the material equivalent to hereditary characters. It was further demonstrated that the expression of genetic character is due to the action of enzymes. The **one-gene, one-enzyme** hypothesis proposed by Beadle and Tatum states that one gene contains the information necessary for one specific enzyme. Today this has been described more accurately - each **structural gene** codes for a specific polypeptide chain. A **sudden change in, or of, a gene (mutation)** leads to a loss of the enzyme or alteration of enzyme that ultimately results in the changes in hereditary characters. Thus the gene is recognised by its mutation and the genetic investigations are the studies of mutants. In eukoryotic system such mutation study is comparatively difficult and complicated as the number of genes in some cases may go to the extents of hundreds of thousand. Bacteria have been identified as ideal tool for the genetic research because :

- (i) they can be propagated rapidly in short duration,
- (ii) genetic homogeneity is maintained in the culture,
- (iii) they are genetically simple organisms having single chromosome,
- (iv) genetic material is easily transferred from one bacterial cell to another that enables to investigate the gene mechanism,
- (v) they require less place and simpler cultural conditions.

GENE : Gene carries character. In both the eukaryotic and prokaryotic cells the molecule that serves as the ultimate agent of chemical control is deoxyribonucleic acid (DNA). A gene is the fraction of DNA molecule that codes for the production of a specific protein or RNA molecule or serves as an operator in controlling the transcription of RNA within an operon unit. An organism's DNA constitutes a catalog of genes known as the **genotype** of the organism. The expression of these genes is referred to as the **phenotype**.

DNA is macromolecule which upon acid hydrolysis yields equimolar amounts of its three components : deoxyribose, phosphoric acid and nitrogen base. The DNA molecule contains four different bases : two **purines** (adenine and guanine) and two **pyrimidines** (cytosine and thymine). In RNA the bases are adenine, cytosine, guanine and uracil. DNA nucleic acids consist of a long chain of nucleotides with alternating pentose and phosphate groups with a base attached to each sugar molecule. Such a nucleotide chain has an orientation displaying phosphate group in the 5' position at one end and a free hydroxyl group in 3' position at the other end. The monomers of RNA are called ribonucleotide because they always contain the five carbon sugar ribose.

Chargaff in 1950 had derived some general law about nucleic acid composition. Adenine and thymine on the one hand and guanine and cytosine on the other always occur in equal amount. The sum of the purines

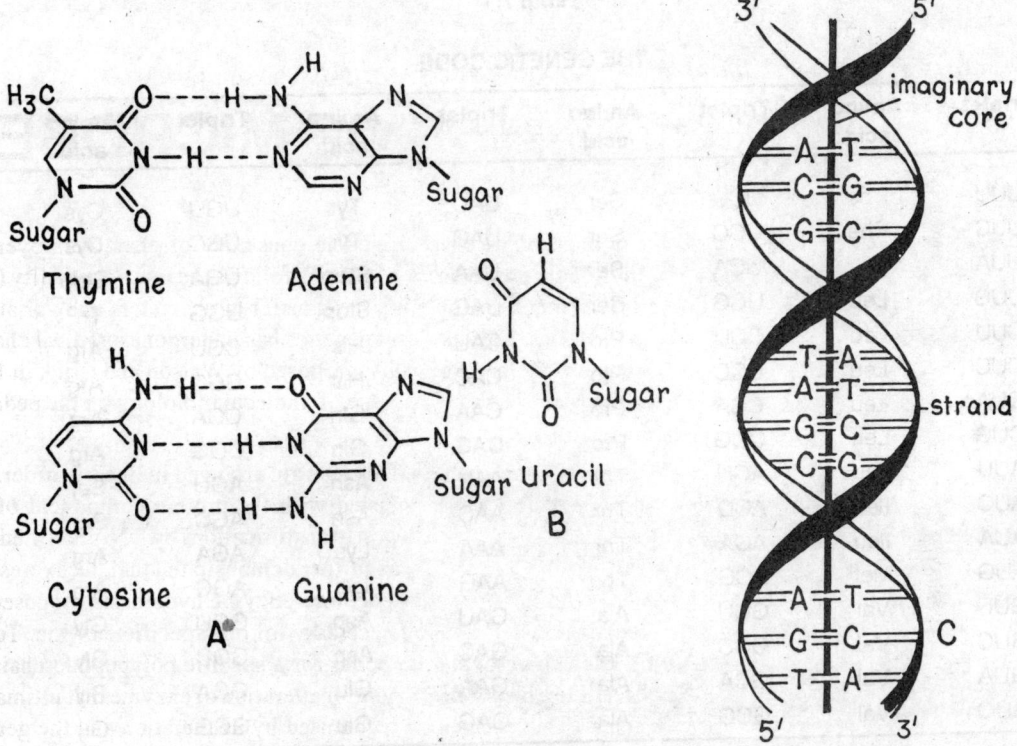

Fig. 7.1   Structure of DNA

A. The pairing of adenine with thymine and guanine with cytosine joined with hydrogen bridge.

B. RNA has uracil in place of thymine.

C. Arrangement of residue in the double strand of DNA molecule.

equals the sum of pyrimidines. The base pair ratio (G+C/A+T) varies from species to species in a wide range but it is constant for a given species. All these informations enabled Watson and Crick to propose a brilliant theory of DNA structure. As per the model, the polynucleotide strands are wound like a double helix around an imaginary axis, the two strands being held together by hydrogen bonds between the internally located bases (Fig. 7.1). Each turn of helix contains about ten base pairs (A-T, G-C). The sequence of the bases in the two strands is therefore complementary whilst the polarity of the strand is in opposite directions i.e., 5'-3' and 3'-5'. The contour length of *E. coli* DNA is about 1-4 nm. The relative mass of 1 μm of double stranded DNA is approximately 2×10^6 or 3000 base pairs {3 kilobase (kb)}.

DNA stores specific genetic information that ultimately determines all the characteristics of an organism. Biological differences between the organisms are primarily due to differences in the information encoded in their chromosomal DNA. The DNA performs three fold biological tasks : **storage of genetic information, inheritence** and **expression of genetic message.**

Task I : The linear sequence of nucleotide bases are arranged in an infinite number of orders. This sequence carries a message written in genetic code (see table 7.1). The genetic information necessary for determining the characteristic of an organism is encoded in the order in which the nucleotides are arranged along the DNA. Any change in the nucleotide base (letter) will change the genetic message thereby changing the characteristics of the organism.

Table 7.1

THE GENETIC CODE

Triplet	Amino acid	Triplet	Amino acid	Triplet	Amino acid	Triplet	Amino acid
UUU	Phe	UCU	Ser	UAU	Tyr	UGU	Cys
UUC	Phe	UCC	Ser	UAC	Tyr	UGC	Cys
UUA	Leu	UCA	Ser	UAA	Stop	UGA	Stop
UUG	Leu	UCG	Ser	UAG	Stop	UGG	TrP
CUU	Leu	CCU	Pro	CAU	His	CGU	Arg
CUC	Leu	CCC	Pro	CAC	His	CGC	Arg
CUA	Leu	CCA	Pro	CAA	Gln	CGA	Arg
CUG	Leu	CCG	Pro	CAG	Gln	COG	Arg
AUU	Ileu	ACU	Thr	AAU	Asn	AGU	Ser
AUC	Ileu	ACC	Thr	AAC	Asn	AGC	Ser
AUA	Ileu	ACA	Thr	AAA	Lys	AGA	Arg
AUG	Met	ACG	Thr	AAG	Lys	AGG	Arg
GUU	Val	GCU	Ala	GAU	Asp	GGU	Gly
GUC	Val	GCC	Ala	GAC	Asp	GGC	Gly
GUA	Val	GCA	Ala	GAA	Glu	GGA	Gly
GUG	Val	GCG	Ala	GAG	Glu	GGG	Gly

Codon UAA, UAG (amber) and UGA are the 'stop' signals (non-sense codon)

Task II : It is the prime duty of the DNA to ensure that all the progeny cells get the necessary genetic information for producing the exact replica of the parent. This is achieved by the **replication of DNA.** Superficially, the increase in DNA i.e., its identical reduplication or replication is very simple. The two strands have merely to separate so that nucleotide building blocks can line up with the complementary bases of each polynucleotide strand and then these can be joined. However, one of the main difficulties encountered in the conception is how the two strands of the double helix separate. According to the X-ray structure diagnosis, DNA is a plectonemical helix. To determine whether unwinding of the helix is necessary assumption, three possible mechanisms of DNA replication were considered (Delbruck and Stent, 1957).

(i) **Conservative mechanism** : This does not involve unwinding of the helix. The parental helix serves as template for the synthesis of two daughter helices so that the daughter double helix consists entirely of new material and the parent helix continues to exist.

(ii) **Dispersive mechanism** : The parental helix is broken down during replication of each half turn by multiple fragmentation. The new synthesis then takes place on the parental fragments with cross-wise annealing. Each polynucleotide strand then contains alternating segments of old and newly synthesised materials.

(iii) **Semi-conservative mechanism** : The parental double helix is unwound and each polynucleotide strand serves as the template for the synthesis of a new complementary strand. The new helix is, therefore, a hybrid consisting of one pre-existing and one newly synthesised strand (Fig.7.2).

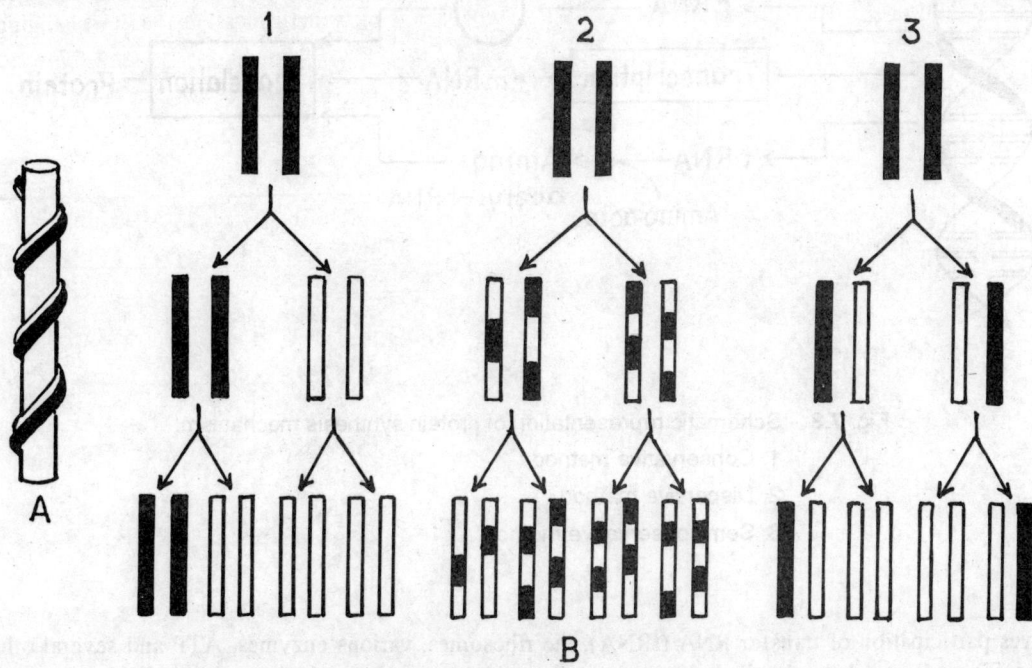

Fig. 7.2   A. Model of plectonemical double helix

B. Three possible mechanisms of DNA replication.

Studies conducted by Maselson and Stahl convincingly proved that out of all the three mechanisms semi-conservative mechanism of DNA replication is the rule. During replication the two strands of the double helix unwind and separate. Each single strand then serves as a template for the synthesis of a new complementary strand. DNA polymerases are involved in this synthesis. They join the correct sequence of new nucleotides to the pre-existing template strand of DNA by base-pairing, thus producing a new polynucleotide strand. Such replication mechanism is found in almost all the organisms. There are, however, considerable difference between the replication of chromosomal plasmid and phage DNA.

Task III : DNA exerts the ultimate control over the cell by dictating the synthesis of specific enzymes and structural proteins. The process by which the expression of genetic message is translated to the formation of protein is known as **protein synthesis**. A brief outline of the mechanism of protein synthesis is depicted in figure 7.3.

DNA is the store house of information but it does not itself serve as template for polypeptide synthesis. The actual synthesis of protein occurs on the ribosomes. The transfer of the information to the sites of protein synthesis is mediated by messenger RNA (mRNA). The synthesis of mRNA takes place on one of the DNA strands (the transcription strand) and by a mechanism similar to that of DNA replication. The amino acids are joined into a polypeptide chain in the sequence determined by the triplet sequence of mRNA in a process that

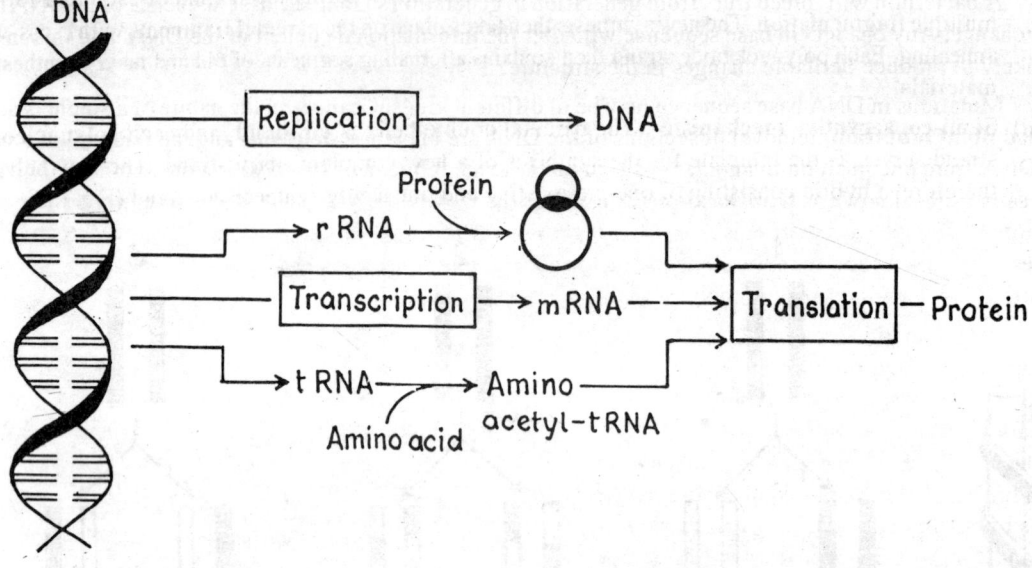

Fig. 7.3   Schematic representation of protein synthesis mechanism.
1. Conservative method
2. Dispersive method
3. Semiconservative method.

involves participation of transfer RNA (tRNA), the ribosomes, various enzymes, ATP and several other factors. To understand the detailed mechanism of protein synthesis, students are advised to consult the relevant literature.

The central dogma of molecular biology states that genetic information flows from DNA to RNA and to protein by the processes of transcription and translation. Some activities of the oncogenic RNA viruses, however, have refuted this belief. Chromosomes of cells infected with these viruses were found to have DNA sequences complementary to the single-strand RNA genome of the virus. This led to the speculation that there may be another kind of polymerase that uses virus RNA as a template to create a complementary strand of DNA. In 1970 an enzyme **reverse transcriptase** (RNA dependent DNA polymerase) was detected in the virion of these oncogenic viruses, having its involvement in the reverse flow of the genetic information. During replication of the retroviruses (*retro* = backward), as these oncogenic viruses are called, the enzyme promotes the assembly of a single stranded DNA molecule along the RNA viral genome. A second strand of DNA is then synthesised, complementary to the first one. It results in double-stranded DNA molecule that may integrate into the host chromosome.

MUTATION : For a long time it was believed that pleomorphism was the rule in the life-cycle of bacteria exhibiting distinct morphological and physiological variations. The view that micro-organisms can mutate and change their hereditary characters was not easily conceived by the microbiologists. However, a number of notable experiments in the early quarter of this century, eventually proved that bacteria do mutate and the mutations occur at random and in non-directed manner. Lederberg's experiments in 1952 provided unassailable proof for selection of mutants. He demonstrated for the first time the technique of **replica plating** which has since become widely established.

A bacterium will 'breed true' from generation to generation so long the base sequence of its DNA does not change. Any change in base sequence will alter the informational content of the DNA and this in turn is likely to produce heritable changes in the structure.

Mutations in DNA base sequence may be of different kinds. Changes in the nature of a single base are called **point mutation**, removal of sections of the DNA are known as **deletions** and the removal of a piece of DNA from one position to another position on the same replicon or to a position on another replicon in the same cell is known as **translocation**. Sometimes the sequence of the DNA is altered either by adding or removing a single base pair. This is known as **frame-shift mutation**. Point mutations themselves are of two types : **transitions** and **transversions**. In the former a pyrimidine is replaced by a pyrimidine or a purine by a purine, while in the later pyrimidines are replaced by purines or vice-versa (Fig.7.3).

Point mutation is caused by a variety of agents known as **mutagens**. Many of these agents are chemical compounds which produce a direct change in the chemical nature of the DNA (e.g., nitrous acid, ethyl-methane sulphonate, mycotoxins etc.). Some ionising radiations are also potent mutagens. They bring about chemical changes in DNA sequence indirectly interacting either with a component of DNA itself or with other molecules in the immediate environment. One of the characteristics of point mutation is that they can revert by a further change in base sequence. Sometimes the DNA returns to its original structure and this is referred to as **back mutation**. The process leads to the production of the original protein and the back mutated cells are indistinguishable from the original parent. A more complex step is the generation of a further **forward mutation** that leads to the appearance of properties similar to but not identical with the original parent. Such a change is called a reversion to the **pseudo-wild-type**. In this case the activity can be reduced or destroyed by the first mutation and then partially restored by the second one.

Bacterial cells are subject to mutations occurring at certain rates without any outside intervention. These are called **spontaneous mutations**. They probably represent accidental error in the assembly of nucleotides during DNA replication. These errors are produced by tautomeric transposition (re-arrangement) of electrons in a base. For example, thymine is normally present in the *oxo* state forming a hydrogen bridge with adenine. However, if thymine were to change to the *enol* form during base pairing that takes place in DNA replication, it would pair with guanine. The new DNA would then contain a GC pair in the position where it would normally have an AT pair.

Usually a mutation is recognised as a sudden phenotypic change in the organism. However, at the molecular level every change may not be expressed phenotypically. In many triplet codons, for example, a change in the third base is without phenotypic consequences. Even a replacement of the first or second base of a triplet does not necessarily have the drastic consequences. Such change is known as **silent mutation**. The frequency of such mutational events can be increased by treating cells with mutagenic agents. This is called **induced mutations** and the resulting mutant cells are called induced mutants.

DNA DAMAGE AND REPAIR

Throughout the course of evolution DNA has been subjected to wear and tear both by endogenous causes (e.g., errors during replication) and exogenous mutagens. This natural burden of mutagens has been increased considerably by the activities of men themselves by polluting the environment with continuous discharge of chemicals and by military and civil use of nuclear fission. Such mutagenic activities lead to various forms of DNA damage. Some major forms of DNA damage are :

Hydrolytic damage (Reaction with water): This includes loss of bases (depurination/depyrimidination) and deamination of exocyclic amino groups.

Adduct formation : Covalent binding of chemicals to DNA with the formation of chemically stable adducts plays a major role in the mode of actions of chemical mutagens. Adducts range in size and complexity from simple alkyl groups (e.g., methyl, ethyl) to aromatic hydrocarbons, aromatic amines, aflatoxins etc.

DNA bases are the main target of attack, the most vulnerable being guanine. Adducts can form link between adjacent bases on the same strand (intra-strand cross-link) and can form inter-strand croos-link between each strand of the duplex. **Mispairing** of bases caused by the presence of an adduct can lead to base pair substitution. Adduct formation can lead to substitution, deletion and addition causing point mutation. However, the primary DNA damage by adduct is not itself a mutation but a **pre-mutational lesion** - a mutation waiting to happen.

Strand breakage : The sugar phosphate DNA backbone can be cut in several ways. Adduct formation may modify the chemical structure of DNA by depurination and hydrolytic cleavage of strand. Physical agents such as X-ray can cause single and double strand breaks.

Electromagnetic radiation - Ionising radiation (X-rays, Y-rays) causes a variety of DNA damage; much of which is mediated by free radicals which induce chemical modifications of DNA bases and strand breakage. Ultraviolet radiation (254 nm) causes adjacent pyrimidines to form dimers (Fig. 7.4) which interfere with DNA replication and whose presence is mutagenic.

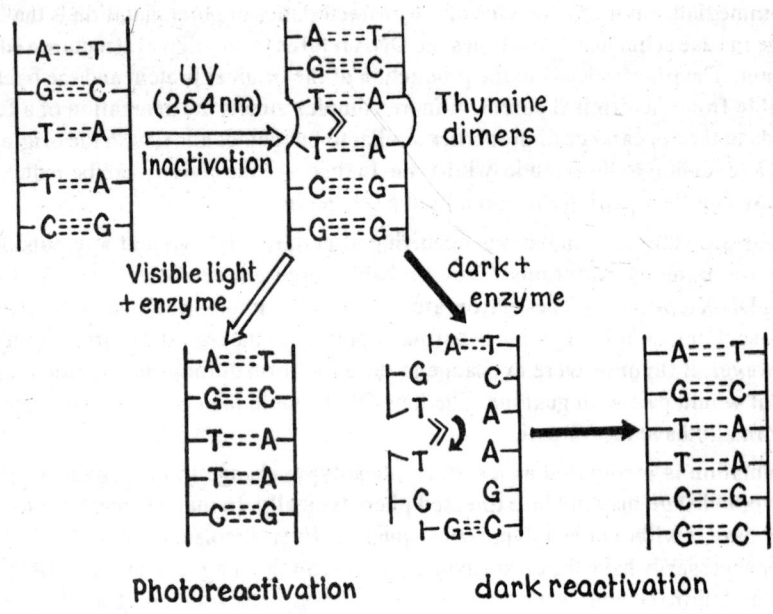

Fig. 7.4   Changes in DNA produced by UV radiation and process of photo - and dark reactivation.

DNA REPAIR

A cell which has sustained DNA damage may respond in several ways, such as :

   (i) it might repair the damage and restore its DNA to the pristine state in which case mutation will not occur, the repair being **error-free**.

  (ii) the cell might die, the possibility of mutation is eliminated.

 (iii) the cell may possess repair mechanisms which allow it to survive and divide despite a burden of pre-mutational lesions.

A number of DNA repair mechanisms have evolved which allow to remove or circumvent the effect of DNA damage.

Error-free repair : This is the best understood mode of DNA repair that operates at several levels of complexity. The simplest is probably **photoreactivation** whereby short-wavelength UV-induced dimers are snipped apart, restoring the DNA to its original state by enzymes which require absorption of light of longer wavelength for their complete activity.

Excision-repair : This is a much more complicated process; the first step may involve enzymes called **glycosylases**. These recognise cut out physically or chemically modified base from the sugar-phosphate DNA chain leaing behind a hole (an ampurinic site 'AP'). Another enzyme (*AP*-**endonuclease**) recognises this hole and cuts the phosphodiester chain at the AP. This cut attracts another type of enzyme (**exonuclease**) which may remove the deoxyribose phosphate and one or a few nucleotides. An alternative first step might be the operation of a damage specific **endonuclease** which recognises the distortion caused by a thymine dimer and slices out the dimer together with a few adjacent nucleotides. At this stage the strand which sustained the damage now contains a gap, opposite its undamaged partner strand which still has its sequence of bases intact. This gap is filled by base pairing by further enzymes (**polymerases**) and the patch of nucleotides is sealed in by yet another enzyme (**ligases**). Thus the coding function of the damaged strain is restored by the process of complementary base pairing, the undamaged strand being the template.

Error-prone repair : This is a strategy which allows cell survival at all costs, even at the risk of incurring a high level of mutation. The essential feature of error-prone repair system is that the normal rules of complementary base pairing are somehow suspended in order to allow DNA replication to proceed even on a damaged template containing pre-mutational lesions. As a result many wrong bases are inserted in the newly synthesised DNA strand. This mispairing leads to a high level of mutation. Recombination of DNA strands during replication of damaged DNA may contribute to the increased level of mutagenicity.

REGULATION OF GENE EXPRESSION

The bacterial chromosome is covalentely closed, that is the ends of poly-nucleotide threads are covalently joined to make a continuous circular molecule. DNA in such a configuration is said to be of CCC form. The piece of DNA able to replicate and segregate independently are known as **replicons**. Each replicon has the ability to synthesise protein or enzyme of its choice. So far the biosynthetic relationship between DNA and protein synthesis is concerned, it was originally believed that every DNA of bacterial cell was always expressed as protein. However, it is now known that only certain regions of DNA carry the necessary information to dictate the amino acids along the polypeptide chain of a single protein and some regions are simply the targets at which effector proteins bind directly to the DNA. Regions that act in such specific way are called **regulatory gene**. These regulatory genes in general control the initiation of transcription and distinguish themselves from genes responsible for specifying protein products, known as the **structural genes**. The structural genes always specify protein products but the regulatory genes may specify a protein product or may not. In practice many enzymes in bacteria contain more than one polypeptide chain and consequently the gene that specifies their synthesis has to contain the appropriate number of sections to code for the individual chain. Such sections are called **cistron** (Fig.7.5).

Regulation of gene expression usually requires the presence of another molecules, other than the protein produced by the molecular gene. This is either the **inducer** or **repressor** depending on its mode of action and process of regulation. The basic philosophy of gene regulation is that the 'resources are conserved by turning off genes for unneeded functions'. For example, many catabolic enzymes are not synthesised unless the substrate is available. The presence of substrate turns on transcription of the appropriate gene, the enzyme is produced and the substrate is utilised. In the absence of the substrate, transcription is turned off. This control

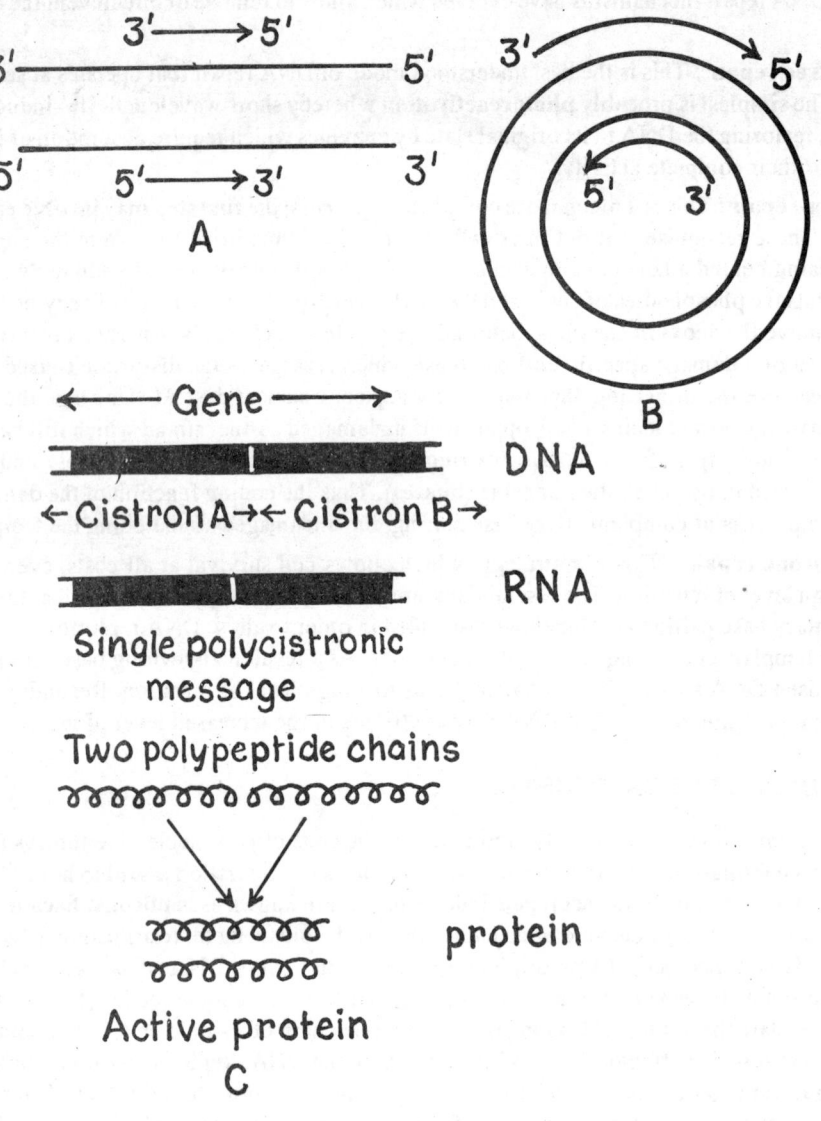

Fig. 7.5 A. Linear double stranded DNA
    B. Covalently closed circular (CCC) DNA
    C. Arrangement of cistron in a gene coding
       for a protein of two polypeptide chain.

of gene expression is **induction** (Fig.7.6). On the other hand if the end product is **abundant, production of** biosynthetic enzyme has to be turned off. This is achieved by the process of gene **repression** (Fig.7.7). Induction and repression are, therefore, the methods regulating enzyme synthesis. **The overall mechanism** that provides this ability to the bacterial cell is called the **operon** (Fig.7.8). The operon is thus a **segment on** the bacterial chromosome that consists : (i) genes for producing enzymes (ii) a regulator region having a promoter and an operator and (iii) a repressor gene. The **promoter** is the site where mRNA polymerase

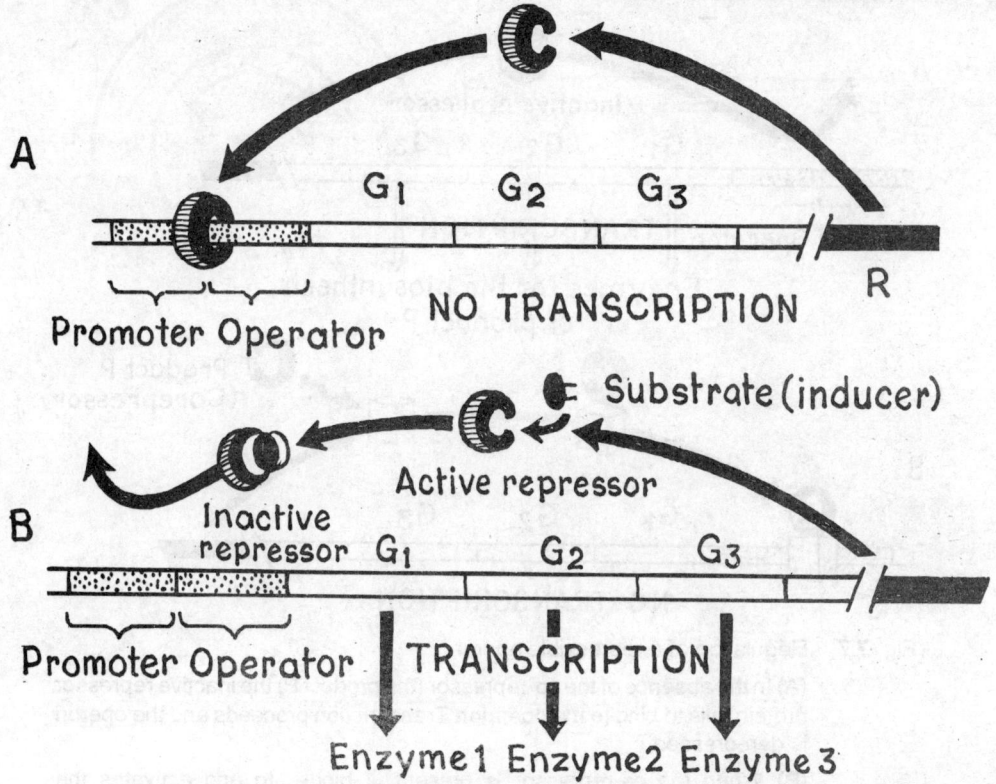

Fig. 7.6   Regulation of an inducible operon

A. In the absence of the inducer (substrate) the repressor protein binds to the operator and blocks transcription by gene by preventing binding of mRNA polymerase. The operon is repressed.

B. When the inducer is present it binds to the repressor and prevents its attachment to operator. Transcription of genes occurs unimpeded.

(transcriptase) binds to DNA and initiates transcription of the genes for enzyme production. The **operator** region lies adjacent to the promoter and is the binding site for a specific repressor protein. When the operator site is occupied by repressor, mRNA polymerase can not bind to the promoter on mRNA and transcription is, therefore, blocked. The **repressor gene** (R gene) is located in another portion of the bacterial chromosome. This gene codes for the production of the repressor protein that attaches to the operator region. It has the ability to specifically bind with effector molecule to determine the operation of operon.

GENETIC TRANSFER IN BACTERIA

Bacteria possess only one set of genes. Unlike eukaryotes where genetic exchange is the result of sexual reproduction, bacteria are capable of donating the genetic information (DNA) to the recipient cells facilitating the latter to acquire new characters. Genes are transferred in bacteria in only one direction - from donor to recipient. In most of the cases a part of the donor DNA is transferred to the recipient cell which aligns with the corresponding segment on the existing chromosome by breakage of the host chromosome and reunion of the free ends with the newly received DNA fragment. Recombination therefore, stably incorporates the new genes into the recipient cell, the **recombinant**.

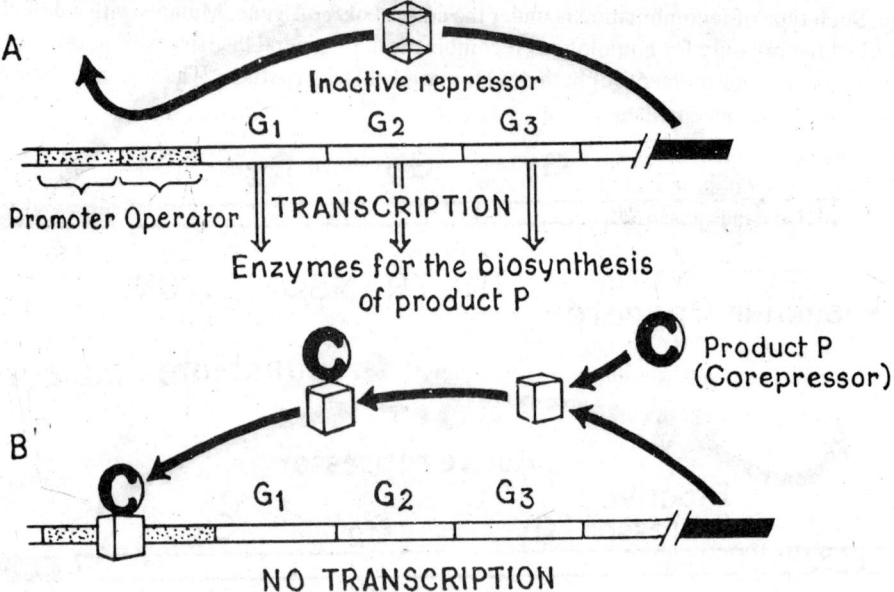

Fig. 7.7   Regulation of a repressible operon

(A) In the absence of the co-repressor (the product P) the inactive repressor protein fails to bind to the operator. Transcription proceeds and the operon is derepressed.

(B) When the co-repressor is present, it binds to and activates the suppressor protein. The repressor-corepressor complex attaches to the operon and blocks transcription.

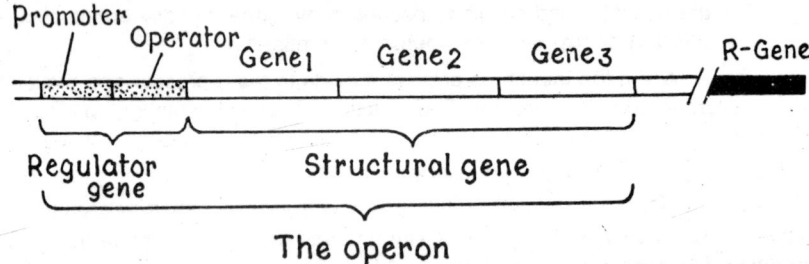

Fig. 7.8   A classical bacterial operon. The promoter and operator segments are located on the end of the operon where transcription is initiated. The R gene produces a repressor protein that binds to the operator and turns off transcription.

GENETIC RECOMBINATION

According to the current state of knowledge donor DNA can be recombined with the recipient chromosome by three possible ways : (i) a general homologous recombination (ii) a site specific recombination and (iii) a non - homologous recombination.

Homologous recombination : This involves the mechanism by which the DNA that has been transferred into the recipient cell recombines with the resident DNA by reciprocal exchange of DNA section.

The recombining DNA partners must have more or less the same base sequence i.e., they exhibit maximal homology. Such type of recombination is under the control of *rec*A gene. Mutants with a defect in this gene (*rec*-) have lost the capacity for homologous recombination. In general the base pairing takes place between unwound single stranded segments of both double stranded DNA partners. The second strand arises by the replication or by repair mechanism.

Site specific recombination : This type of recombination can occur in *rec*- mutants. It consists of integration of a small, double stranded DNA segment at a specific site of large-double stranded recipient DNA. The small partner loses its autonomy by the integration. The model example for site-specific recombination is the integration of the bacteriophage to the host chromosome. It was earlier thought that the contact was due to homology between the base sequence but now it is known that the homology of these segments is only slight. The process actually involves a phage-coded protein **integrase** that catalyses a break at a specific site (*att*B) in DNA and another break at a specific site of the host DNA (*att*) and the subsequent crossed reunion of the phage and the host genomes.

Non-homologous recombination : This represents an integrative form of recombination similar to that in the site-specific process involving addition of DNA rather than exchange. This is also largely achieved by the involvement of **insertion sequence** (IS elements) or **transposon** (Tn elements).

IS element : These are short DNA sequence (800-1400 base pairs) found in main host chromosome as well as in plasmid. They do not code for any recognisable phenotypic character but play an important role in mediating the integration of plasmids into the host chromosome. IS elements are transposable i.e., they can move from one position to another within a chromosome or to a different chromosome. It is assumed that IS elements determine the re-orientation and joining of genetic material.

Tn element : Transposons are a group of more complex transposable elements that behave genetically much like the IS elements. They are usually 2000 base pairs in length and carry one or more genes unrelated to their transposability. These additional genes provide resistance to one or more antibiotics. The most striking structural feature of Tn is that they all have the same or very similar sequences repeated at the both ends (so called terminal repeats). These terminal repeats play a key role in the transposability of genetic elements. Tn elements are more easily recognised than the IS element. The resistance genes in a transposon are flanked by two DNA segments with repeated base sequences. The arrangements of these flanking DNA sequences can be seen in the electron microscopic heteroduplex analysis (Fig.7.9).

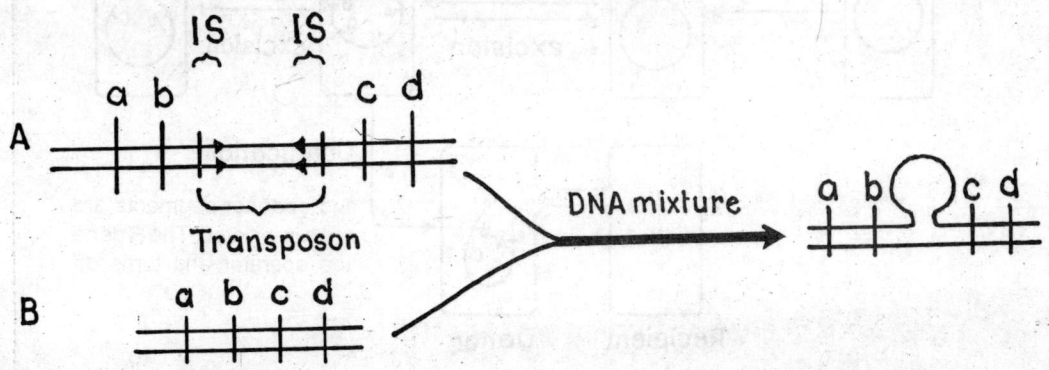

Fig. 7.9   Heteroduplex analysis for the recognition of transposon. Transposon forms a 'bubble'

A. Transposon carrying bacterium

B. Wild-type bacterium (IS = Insertion sequence).

In bacteria transfer of genetic characters is achieved by three ways : **congugation, transduction** and **transformation**. The three processes differ only in the manner in which the DNA is transferred.

Conjugation : Conjugational transfer of DNA requires cell to cell contact by the participating bacteria. The process is often referred to as bacterial mating. During conjugation, the DNA passes from the donor to the recipient in a protected form. Unequivocal proof of the transfer of genetic material by direct contact was first of all obtained by Lederberg and Tatum in 1946.

In practice bacterial plasmids are by far the most common form of DNA to be passed from cell to cell by conjugation. It has been observed that the donor state in bacteria is dependent on the presence of transferable DNA element in the cell known as **sex factor** or **F-factor** (F for fertility). It is a type of plasmid that contains the genes responsible for conjugation. The process is accomplished through a special structure on the cell surface, the F-pili. Cells that lack F factor (F⁻ cells) are the recipient ones. The F factor is supposed to be transferred during conjugation to convert the recipient cell to a potential donor cell. However, transfer of the F factor does not necessarily involve the transfer of chromosomal characters. Only a few bacteria in a F⁺ population are able to transfer chromosomal DNA. These are cells in which the F factor is integrated into the bacterial chromosome. When clones of such donor cells are used in conjugation experiment, the yield of recombinant is about 1000-fold higher than the ordinary F⁺ strains. Such cells are called **Hfr** (high frequency recombinant) cells.

The integration of the F factor into the bacterial chromosome is reversible. The F factor can be excised from the chromosome so that the Hfr cell reverts to a F⁺ cell. This excision occurs at about the same frequency as integration and the site of break is usually the site of integration. However, in some cases, the break occurs at a neighbouring sites so that on excision a neighbouring segment of DNA remains attached to the **F** factor. Such a **F** factor, containing a small piece of chromosomal DNA is called an **F'** factor and the cell is called a **primary F' cell** (Fig-7.10). The DNA integrated into the F factor can be transferred from the F¹ donor cell to an F⁻ recipient cell with the same frequency as the F factor, from F⁺ strains to F⁻ strains. Transfer of F¹ factor from a primary F' cell in which it originated to a normal F⁻ cell results in a **secondary F' cell**. In this cell the segment of the bacterial chromosome remains in diploid state.

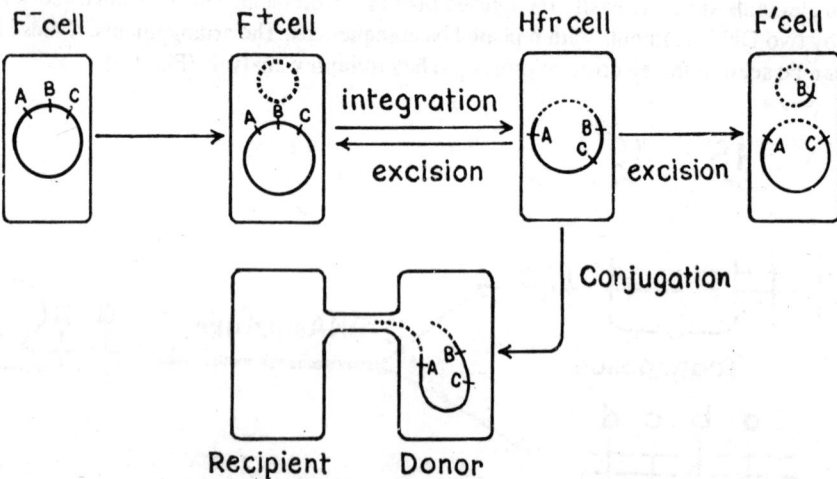

Fig. 7.10 Conjugation in *Escherichia coli* F⁻ cells are recipients. By conjugation with an F⁺ or Hfr strain they can receive the F factor and also become F⁺. When F factor is integrated into the bacterial chromosome it makes the cell Hfr. Incorrect excision of the F factor from chromosomal DNA gives rise to a F factor containing, short chromosomal DNA. This forms F' cell.

Plasmids : Plasmids are autonomous genetic elements that can be found in almost every known type of bacterial cell. They have an independent self-replicating existence quite distinct from the main chromosome of the cell. They are highly variable in size ranging from those carrying no more than three genes to those large enough to carry several hundred genes. Certain bacterial cells are known to harbour as many as 11 different plasmids in addition to the main chromosome. Most of the plasmids are not essential for growth of the bacteria under normal conditions. However, in many cases they are essential, such as in the presence of antibiotics these help tide over the unfavourable environment. Many plasmids can integrate into or out of the main chromosome much like temperate phage λ. Plasmids and temperate phage with such integration properties are known as **episomes**. Strictly speaking the chromosome of mitochondria and chloroplasts in eukaryotes also fit in the definition of plasmid owing to their self replicating autonomous characters; the only remarkable difference is that plasmids are not organised in organelles.

On the basis of genetic information that are carried, plasmids may be termed as : **R-plasmid** (antibiotic resistance), **F and F' plasmid** (sex plasmid or conjugative fertility-factor) or **Col-plasmid** (specify protein colicin). Plasmids are also known to code for bacteriocin other than colicin. For example, plasmids code for vibriocins that kill sensitive *Vibrio cholerae* cells.

Some plasmids enable host cells to conjugate with other cells. Based on the fact whether or not they mediate conjugative self-transfer, plasmids can be grouped into two major forms :

(i) **Conjugative or transmissible plasmids** : that mediate transfer of DNA by conjugation e.g., all F and F' plasmids, many R plasmids and some Col-plasmids are conjugative. The conjugative nature of many R plasmids has major significance in the rapid spread of antibiotic and drug resistance-gene.

(ii) **Non-conjugative or non-transmissible plasmid** : that do not mediate transfer of DNA by conjugation e.g., some R plasmid and Col - plasmid.

R-plasmid : R factors contain genes that make the host bacterium resistant to antibiotics and drugs e.g., sulfonamide, streptomycin, chloramphenicol, tetracyclin etc. Some resistance factors determine resistance to as many as eight antibiotics. Other confer resistance to toxic heavy metals such as mercury, nickel, cadmium or cobalt. All conjugative **R** plasmids have at least two components : one segment carrying a set of gene involved in conjugative DNA transfer (similar to *tra* gene of F plasmid) and a second segment carrying the antibiotic resistance gene. The former is called **RTF** (Resistance Transfer Factor) and the latter is known as **R-determinant**. The resistance transfer factor (RTF) includes all those genes that are responsible for the transfer of R factor from cell to cell, usually by conjugation. The RTF region, in fact, is found to have a large degree of structural homology with the F factor of *E. coli*. The R factors are characterised by a wide host spectrum; they can be transferred between several bacterial genera.

The mechanism of antibiotic resistance that is mediated by resistance factors may be different from that determined by chromosomal mutation. The drug-resistance gene in R plasmids are symbolised differently than the comparable gene in chromosome. The chromosomal streptomycin-resistance gene is denoted *Str*-r- while the plasmid borne gene is written as *Sm*. Likewise ampicillin is denoted as *Ap*; chloramphenicol, *Cm*; kanamycin *Km* and tetracyclin, *Tc* etc. The basic difference between chromosomal and plasmid resistance is typically the result of an **alteration in a ribosomal protein** whereas the plasmid gene typically dictate the synthesis of enzymes that inactivate the antibiotics as they enter the cell. For example the *Cm* gene directs the synthesis of *chloramphenicol acetyl transferase,* an enzyme that adds in inactivating acetyl group to the chloramphenicol molecule. Similarly penicillin is inactivated by the enzyme *penicillinase*. R factors are of great importance for chemotherapy because they can undergo genetic recombination through whole bacterial population producing new recombinants.

Since plasmids have the ability to replicate independently and to combine with other DNA, they are useful in **genetic engineering** . Some properties coded by naturally occurring plasmids are listed in table 7.2. Plasmids can carry gene for specific biochemical properties that may be of selective advantage under certain conditions. Genes for enzyme involved in the degradation of camphor, salicylic acid, naphthalene, octane and many other unusual substrates can be localised on plasmid. A tumor inducing plasmid T_1 carries a DNA sequence that transforms cells of dicotyledons (tobacco, sunflower, carrot, tomato etc) to tumor cells which

Table 7.2
NATURALLY OCCURRING PLASMIDS AND THEIR PROPERTIES.

Plasmids		Properties
F, R 1, Col 1	-	Fertility-ability to transfer genetic material by conjugation.
Clo DF 1	-	Production of bacteriocin.
R 222	-	Antibiotic resistance.
R 1258, R6	-	Heavy metal resistance.
Ent	-	Enterotoxin.
T_1-plasmid	-	Tumorigenicity in plants.

is associated with crown gall disease. This disease is induced by a bacterium *Agrobacterium tumefaciens*. Genes carried by plasmid integrate with plant cells and code for enzymes that promote continuous and uncontrolled tumor growth. The list of bacterial properties that can be coded by plasmid is becoming long. It includes nitrogen fixation, formation of root nodules, production of indole acetic acid, sugar uptake, formation of diacetyl and of hydrogenase. It is also believed that plasmids have played a decisive role in the evolution of prokaryotes.

Transduction : Transduction is the transfer of DNA from a donor cell to a recipient cell by bacteriophage. Two kinds of transduction can be distinguished : a non-specific transduction which can transfer any part of the host DNA, and a specific transduction which is restricted to the transfer of specific DNA segments. In non-specific transduction the host DNA segment is integrated into the virus particles either in addition to or in place of some of the phage genome. In specific transduction some of the phage genes are replaced by the host genes. In both the cases the transducing phages are usually defective in some respect; for example, they often lose the ability to lyse host cells.

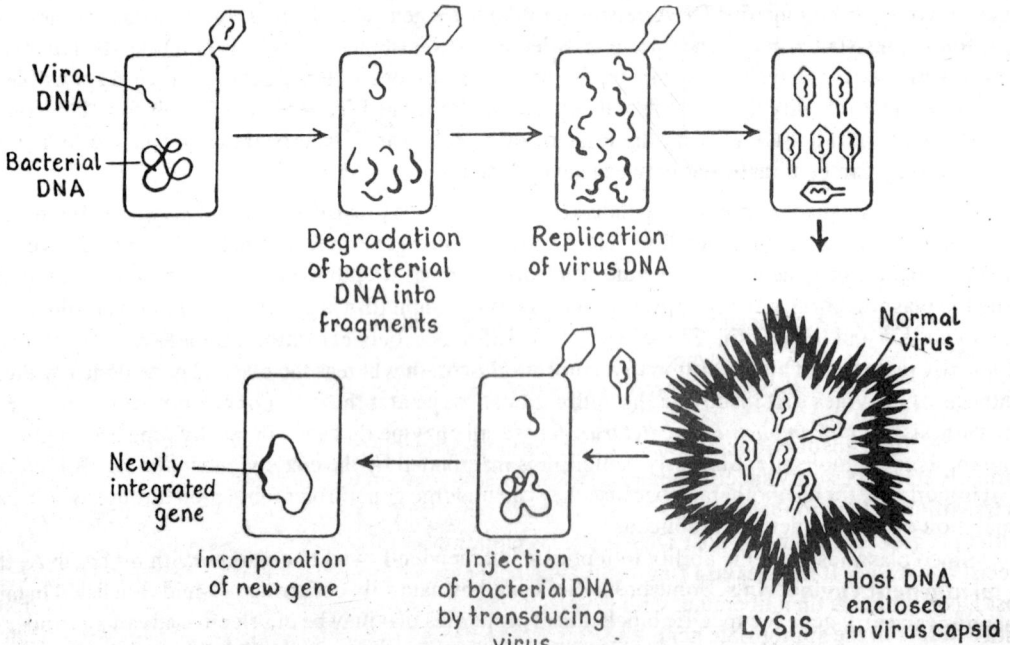

Fig. 7.11 Non specific (generalised) transduction.

Non-specific transduction : It is mediated by the prophages that remain in the cytoplasm as plasmid, that are not attacked to the chromosome. During the lytic multiplication of the phage inside the doner strain, the host DNA is broken into fragments and become incorporated in the capsids accidently in place of phage DNA. A phage lysate thus contains a mixture of normal and defective phages. Usually infection of the recipient cell by normal phage leads to lysis but infection with a defective transducing phage results in the recombination of phage DNA with the chromosome of the recipient. Homologous DNA segments are thus exchanged and this gives rise to complementation of a defective gene in the recipient cell (Fig. 7.11). This is a random occurrence that may involve any of the bacterial genes, hence the name generalised transduction. Such transduction is very much evident in *Salmonella* phage P 22, coliphage P 1 and *Bacillus subtilis* phage PBS 1.

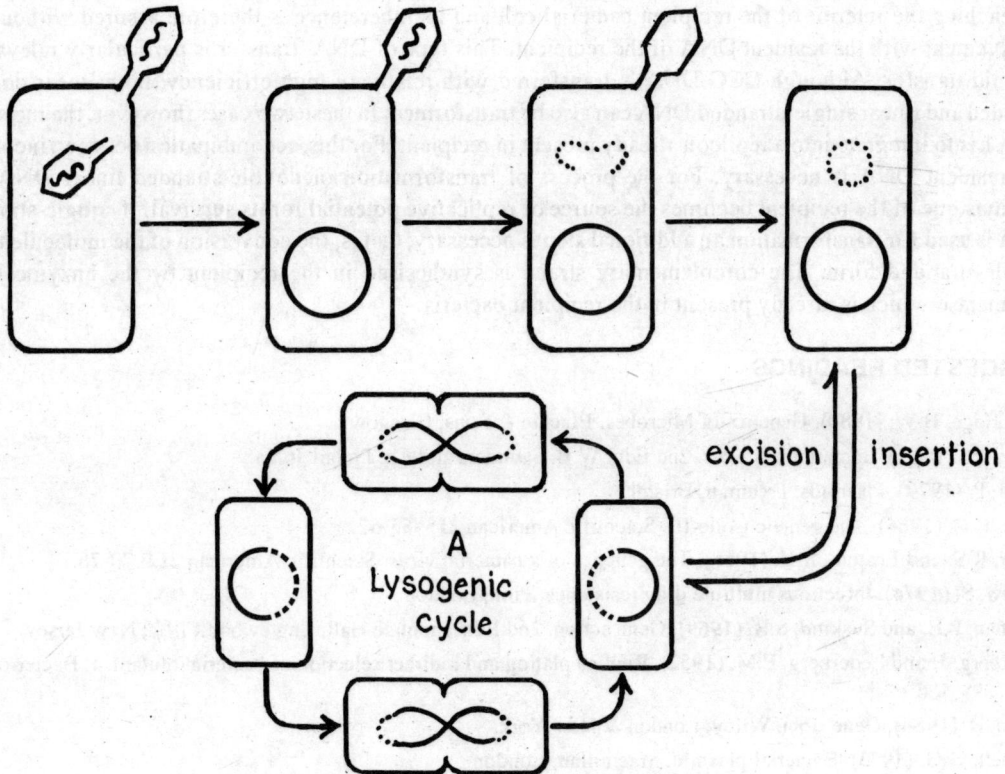

Fig. 7.12 Specialised transduction in phage lambda.

A. The lysogenised bacterium can go on dividing indefinitely until the excision of the phage DNA from host chromosome is achieved.

Specific transduction : The best known example of specific transduction is phage (Fig.7.12). Lambda normally tranduces only specific gene of the *gal* and *bio* operons. It integrates into the host genome during its transition to the prophage state, site -specifically between the *gal* and *bio* operons. During the separation of the phage DNA from the host DNA (by UV radiation or other mutagens) the phage DNA may not be precisely excised. It may leave a fragment in the host chromosome and acquire a neighbouring piece of the host DNA which is then liberated with the phage DNA. When a transducing phage infects a recipient cell with a defect in the appropriate gene such as *gal* the intact transduced gene can exchange with the defective host gene. All resulting recombinants or transductants are, then *gal*⁺.

Transformation : Transformation was first discovered by Griffith in 1928 who observed that genes can be transferred even without cellular contact or vectors. The DNA involved in this transfer process is liberated from the donor bacterium by lysis and then taken up by the recipient bacterium. This is the most important kind of gene transfer in bacteria. The potential for transformation is apparently restricted to bacteria that are able to take up high molecular weight, double-stranded, intact DNA. This ability is defined as **competence**. Although competent bacteria can take up any DNA, only the DNA of closely related species can lead to recombination mainly by exchange of homologous DNA fragment. The potential to transfer genetic characters by isolated DNA has since been established in *Haemophilus influenzae, Neisseria, Rhizobium, Bacillus subtilis, Acinetobacter, Escherichia coli, Pseudomonas* etc.

The highest frequency of transformation is normally obtained when the transforming DNA is double-stranded and circular. Under these circumstances the incoming DNA can usually form a replicon immediately on reaching the interior of the recipient bacterial cell and its inheretence is therefore assured without any involvement with the resident DNA of the recipient. This type of DNA transfer is particularly relevant to plasmid transfer. Although CCC DNA is transferred with relatively high efficiency, both linear double-stranded and linear single-stranded DNA can also be transformed. In these two cases, however, the incoming DNA has to integrate into a replicon already present in recipient. For this, recombination between incoming and resident DNA is necessary. For the process of transformation of double-stranded linear DNA, the chromosome of the recipient becomes the source of replicative potential for its survival. If single-stranded DNA is used for transformation an additional step is necessary, that is, the conversion of the molecule to the double-stranded form. The complementary strand is synthesised in the recipient by the enzyme DNA polymerase which is already present in the recipient bacteria.

SUGGESTED READINGS

Bainbridge, B.W. (1980). Genetics of Microbes. Blackie & Sons, Glasgow.

Braun, W. (1965). Bacterial Genetics. 2nd Edn. W.B. Saunders and Co. Philadelphia.

Broda, P. (1979). Plasmids. Freeman, Bristol.

Crick, H.C. (1966). The genetic Code-III. Scientific American **215**: 55-62.

Edger, R.S. and Epstein, R.H. (1965). The genetics of a bacterial virus. Scientific American **212**: 70-78.

Falkow, S. (1974). Infectious multiple drug resistance. Pion, London.

Hartman, P.E. and Suskind, S.R. (1969). Gene action. 2nd Edn. Prentice Hall, Englewood Cliffs, New Jersey.

Lederberg, J. and Lederberg, E.M. (1952). Replica plating and indirect selection of bacterial mutant. J. Bacteriol, **63**: 399-406.

Lewin, B. (1984). Gene. John Wiley, London & New York.

Meynell, G.G. (1973). Bacterial plasmid. Macmillan, London.

8

MICROBES AS RESOURCES

Man's concern to microbes was primarily restricted to their role in diseases. Since the emergence of Louis Pasteur in the horizon of microbiology the entire concept of micro-organisms has undergone a radical change and today the microbes are regarded as ideal tools for many biological transformations. Recent advances in molecular biology have opened new frontiers in the field of industry, agriculture, medicine and food production. Sophisticated techniques are providing new and exciting ways of tapping the technological potentialities of microorganisms in different arenas. Yet, the applied fields of microbiology are still dominated by traditional applications of microbial activities viz., industrial fermentation, production of alcoholic beverage, food, antibiotics and vaccines etc. In this chapter we will discuss the conventional applications of microbes in conjunction with some newer possibilities of their exploration as resources for humans.

MICROBES IN FOOD PRODUCTION

Foods serve as excellent media for the growth and regeneration of micro-organisms. As natural contaminants these microbes, at one hand contribute significantly in food spoilage, on the other they influence aptly raising the food value of the commodities. Besides being directly involved in the production of commercial materials like dairy products (cheese, butter, yogurt etc.), non-dairy food (vegetables, soy sauce, bakeries), alcoholic beverage etc., microbes are exploited in other ways as well. Cultivation of microbial cell as protein supplements in feed (**Single Cell Protein - SCP**), as food additive for flavour, consistency and nutrition as well as production of enzymes or other metabolic products enhancing the quality of flavouring are some of the important technological advancements in food industry.

SINGLE CELL PROTEIN - (SCP)

For a long time it was believed that all the proteins had the same nutritional value. It is now, however, realised that proteins vary considerably in their amino acid constituents. The emphasis has, therefore, shifted from the whole protein molecule to their amino acid components of dietry significance. These nutritionally rich amino acids are regarded as the essential amino acids. For humans, out of nearly twenty known amino acids only eight are considered to be essential viz; isoleucine, leucine, lysine, methionine, phenylalanine, threonine, tryptophane and valine. Recently histidine and arginine have also been added to the list and are considered to be semi-indispensible amino acids. Lack of these essential amino acids in the diet results in malnutrition and sometimes many serious diseases. **Kwashiorker** is a fatal disease in Japan caused due to deficiency of lysine in the diet. Children of Africa and Central Asia are known to suffer from **marasmus, a** deficiency due to lysine and methionine.

World's demand of protein is approximately 180 million tonne, out of which only 82 million tonne is supplied from animal and vegetable sources. Out of this amount of supplied protein only 40 million tonne is supplemented from cereals and legumes and rest is met by animal protein. However, in cereals or vegetables this gross quantitation of protein lacks one or more essential amino acids. For example, wheat lacks lysine; maize is deficient in lysine and tryptophane; rice in lysine and threonine both whereas legumes are deficient in methionine, lysine and tryptophane. In order to alleviate malnutrition and meet the requirement of protein demand quantitatively as well as qualitatively attempts are being made to tap some cheap and alternative sources of natural protein. Microbial protein, popularly known as single cell protein, is an ideal substitute in the hand of mankind.

Gray (1962) has depicted an illustrative chart (Fig. 8.1) of the possible exploitation of microbial protein

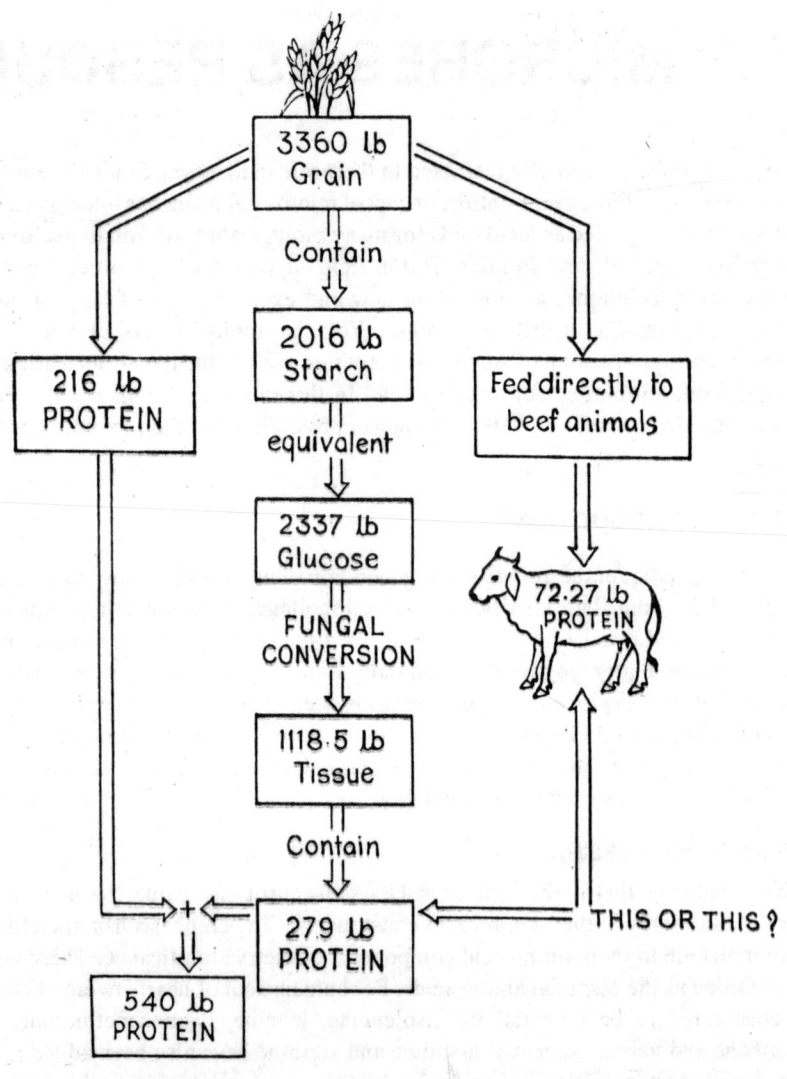

Fig. 8.1   A comparative picture of animal protein and microbial protein source showing microbial protein having definite edge over animal protein.

vis-a-vis the animal protein which unequivocally presents a preferential edge of the former over the latter. Single cell proteins have some obvious advantages over the traditional protein sources such as (i) microbes can grow rapidly in inexpensive indigenous material (ii) have simple nutritional requirement (iii) are easy to harvest (iv) have easy scope of genetic manipulation for better yield (v) are rich in protein, fat and carbohydrate (vi) are very good source of vitamins also and above all (vii) are independent to climatic changes.

Algae, filamentous fungi, yeast and bacteria are the well exploited organisms for the production of SCP. Some important members of these groups known for the production of SCP are listed in table 8.1.

Table 8.1

MICRO-ORGANISMS PRODUCING SINGLE CELL PROTEIN

ALGAE	*Chlorella, Scenedesmus, Laminaria, Spirulina*
FUNGI	*Fusarium, Aspergillus, Penicillium, Rhizopus*
BACTERIA	*Lactobacillus fermentans, Nocardia, Alcaligens viscosus, Escherichia coli*
YEAST	*Candida, Torulopsis, Kluyveromyces fragalis, Hansenula polymorpha*

The protein content of algae ranges from 40 to 60 percent that contains most of the essential amino acids. Lysine and threonine are fairly rich in algae. However, these are poor source for methionine. Other essential amino acids are present in good concentration. Since algae utilise solar energy more efficiently and can grow in arid and waste land, these have been suitably tapped for SCP. Besides accumulating good amount of carbohydrate and lipid, many of the algae are rich source of vitamin C and vitamin B complex.

Amongst micro-organisms bacterial cell contains highest amount of protein content that ranges from 78-87%. Besides being rich in lysine other essential amino acids viz; tryptophane, threonine and methionine are also found in good concentration. Bacterial cells have very rapid growth rate in comparison to any eukaryote or prokaryote and have the ability to grow on hydrocarbon. These are also known to be good source of vitamins.

Fungi do contain 19-47% protein contents and are fairly rich in methionine. Riboflavin and vitamin B_{12} have also been extracted in good amount from many fungi. These can grow in crude raw materials and are easily handled for the extraction of fungal protein.

The most digestible form of protein is met in yeast that contains 45-55% of microbial protein. It is fairly rich in most of the essential amino acids except methionine and is a very good source of riboflavin and pentathenic acid. Having little or no adverse effect on animal system, yeast protein is extensively exploited source of SCP.

In spite of being a source of natural protein these micro-organisms have not yet been able to replace the traditional animal and vegetable proteins owing to certain limitations. Gastrointestinal disturbances are the common experiences of the ingestion of microbial protein. Presence of high nucleic acid contents in the cell quite often leads to kidney stone formation. Degradation of nucleic acids generates uric acid that accumulates in joints to cause gout. Besides, presence of endotoxins in Gram-negative bacteria and mycotoxins have added extra-detrimental consequences of the SCP. Above all, palatibility of SCP has been a major constraint for its acceptance by the common mass. However, with the development of newer techniques in the field of nutrition, manipulating genetic traits of organisms have helped in reducing the constraints to a great extent and the possibilities of utilisation of SCP seem a reality.

FOOD ADDITIVES AND MICROBIAL PRODUCTS

Many vitamins, amino acids, nucleotides and enzymes that are commercially valuable to food industry are readily obtained from microbial cultures. These ingredients contribute significantly in enhancing the flavour and taste of the finished food products. For examples, some mutants of *Corynebacterium glutamicum* produce lysine that is used for supplementing lysine deficient plant proteins in animal feed. Production of baker's yeast is perhaps the largest domestic use of a micro-organism that produces desired aroma and texture to various edibles. Flavour enhancing salt, **monosodium glutamate** (MSG) is commercially produced from glutamic acid, a microbial product of *Brevibacterium* sp. Some propionibacteria can synthesise huge quantities of vitamin B_{12} in culture. Riboflavin and precursors of vitamin A and C are also the microbial products that add the nutritive values of the food. Besides, many enzymes instrumental in food production are also derived from microbes (see Tabel 8.2).

TABLE -8.2
ENZYMES PRODUCED BY MICROBES AND THEIR USES

Enzymes	Microbe	Application
Amylase	*Aspergillus,* *Bacillus subtilis*	Brewing, Syrup production
Diacetyl reductase	*Enterobacter* *aerogens*	Fermenting certain off-flavour in beer and fruit juice
Glucose oxidase	*Aspergillus niger*	Prevention of browning in dried egg.
Glucose isomerase	*Bacillus, Arthrobacter*	Preparation of sweet syrup
Invertase	*Saccharomyces cerevisiae*	Manufacture of candies
Lactase	*Kluyveromyces* fragalis	Reduction of lactose in milk, prevention of lactose crystallisation in ice-cream
Pectinase	*Aspergillus*	Preparation of concentrated fruit juice
Protease	*Bacillus, Aspergillus*	Tendering of meat
Renin	*Endothia, Mucor*	Production of cheese by coagulation of caesin

Recently plasmid DNA studies of the existing lactic acid bacterial strains have opened up possibilities of using DNA technology. With the help of this technique it is feasible to screen bacteria which produce bitter taste or are susceptible to phage attack and produce other undesirable flavours.

MICROBES IN INDUSTRY

Microbes are the living chemical factroy. They can manufacture many industrially important substances that are not readily or economically obtained from other sources. The basic principle of industrial microbiology is to maximise the production of economically valuable products by employing high yielding strains of micro-organisms capable of utilising the inexpensive raw materials or crude substrates. The combination of mutagenesis to obtain the desired mutant strains and manipulation of cultural environment is the key to successful exploitation of microbes in industry as is evidenced by the 10,000 fold increase in today's penicillin production over that of Fleming's original strain.

Pre-requisites of microbe based industries

In order to explore the microbial capabilities fully in industry certain preconditions are essential such as

(i) *Purity and nature of strain* - It must be ascertained whether absolutely pure culture be used or mere predominance of one organism is sufficient to achieve the desired result in terms of sufficient yield of the product.

(ii) *Medium or raw materials* - The medium should be capable of supporting luxuriant growth of the organism and it must be constantly available and profitable. Necessity of preliminary treatment is also important e.g., liming of distillery waste, molasses, digestion of saw dust of fibre wilt etc.

(iii) *Nature of the process* - Any time consuming process e.g., aging, ripening etc. has adverse commercial effect. The more complicated cultural processess like heating, dilution, digestion etc. are likely to escalate the cost of production.

(iv) *Preliminary experimentation* - Any process developed in the experimental laboratory must next prove its worth in the factory. Practice of small sized model or 'pilot' project is beneficial.

During past two decades genetic recombination technology has created new frontiers in industrial microbiology. Recombination results in a single hybrid cell with combined characteristics. The technique is valuable for incorporating into a single organism distinct desirable traits; for example, fusing a cell that promotes high yields with the one that grows rapidly. Gene amplification increases the number of copies of a gene in the cell thereby increasing the amount of corresponding products.

INDUSTRIAL FERMENTATION

Fermentation is actually any energy yielding process in which organic compound (e.g., glucose) is converted to lactic acid or ethyl alcohol through anaerobic process. In industry fermentation is, however, conceived to convey some broader meaning i.e., any chemical process (aerobic or anaerobic) that is catalysed by micro-organisms. Microbial products of industrial importance fall into three broad categories : (i) primary metabolites, (ii) secondary metabolites and (iii) enzymes. Primary metabolites are compounds 'that are either intermediate or end-products of biochemical pathways essential for the growth of the microbes. Sugar, amino acids, vitamins, nucleotides, organic acids, alcohol etc. are the examples of primary metabolites. Secondary metabolites are not essential for microbial growth and in many instances their role is unknown. These are, however, believed to protect the cells against adverse conditions during the rest phase. Secondary metabolites are usually low molecular weight compounds. Many antibiotics and pharmaceutical products are secondary metabolites. Some enzymes and primary metabolites of industrial utility, that are produced by microbes, are listed in Table -8.3. Few common commercial products of microbial origin are discussed below :

Table 8.3

SOME INDUSTRIAL PRODUCTS OBTAINED FROM MICROBES

Products	Micro-organisms	Application
Acetone butanol	*Clostridium acetobutylicum*	Solvent chemical
Cellulose	*Acetobacter xylinum*	Filter, fibre-production
Citric acid	*Aspergillus niger*	Food acid, cosmetics
Dextrans	*Leuconostoc mesenteroides*	Blood plasma expander, adsorbant
Dihydroxy acetone	*Gluconobacter suboxydans*	Fine chemicals
Glutamic acid	*Brevibacterium* sp.	Food additive
Itaconic acid	*Aspergillus terreus*	Plastic industry
Lactic acid	*Lactobacillus delbrueckii*	Food product, textile, laundry
Lipase	*Rhizopus, Saccharomyces*	Degreasing wool digestive aid
Protease	*Bacillus, Streptomyces*	Detergent
Vinegar	*Acetobacter* sp.	Preservative medicine
Xanthans	*Xanthomonas campestris*	Drilling muds, stabilizers and emulsifier

Ethyl alcohol

Production of ethyl alcohol is achieved by fermentation of any carbohydrate containing fermentable sugars. Selected strains of *Saccharomyces cerevisiae*.are commonly used as inoculants to molasses or other fermentable substrates. The yeast, *S. cerevisiae* is fairly tolerent to the high level of ethyl alcohol that is eventually produced by fermentation. Two basic kinds of fermentation technologies are employed.

(i) **Batch fermentation** - A tank or fermenting vat is filled with the nutritive substrate called **mash**. Fermenter's environment (pH 4.5, anaerobic, 25°C/48 hrs) and microbial growth during the various phases of fermentation are maintained. Several auxillary vats are used for growing the inoculum or 'seed' that is inoculated at the rate of 2-5% of the volume of mash. When the fermentation is complete, mash is removed, vat cleaned and fresh batch is filled.

(ii) **Continuous fermentation** - Mash is fed into a fermenter at a fixed continuous rate to replace nutrients. At the same time portions of the fermentating materials are withdrawn from the vessel. Microorganisms act on the substrate as it flows through the fermentation unit.

In the recent past two phase system has found its way in fermentation industry - the first phase involves the production of ethyl alcohol from molasses by the activity of *Saccharomyces cerevisiae,*and the second phase comprises the conversion of ethyl alcohol to acetic acid by the activity of *Acetobacter* sp.

These days ethanol is produced more economically by chemical synthesis from petroleum products. However, improved strains of microbes are able to compete with the chemical process as the cost of petroleum has increased substantially in the recent past and DNA recombinant technology has produced better microbial strains which are capable of utilising cheap organic wastes. In figure-8.2 an outline of utilisation of wood products for the production of ethyl alcohol is depicted.

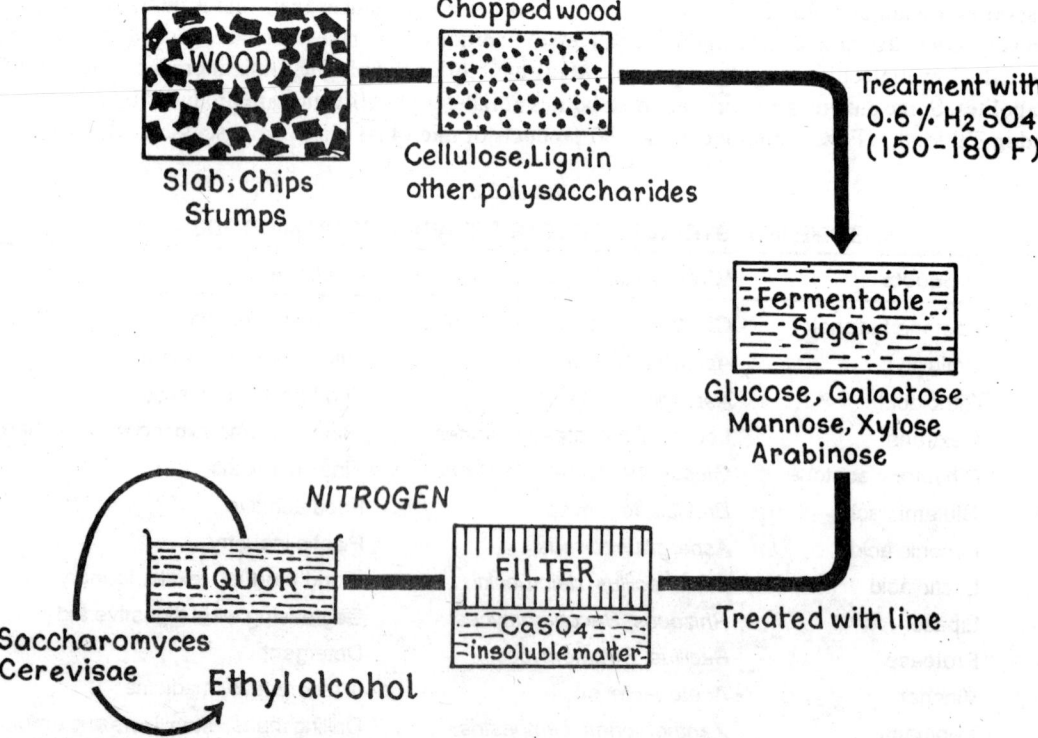

Fig. 8.2 Utilisation of wood products for ethyl alcohol preparation.

Vinegar

Like alcohol vinegar is also one of the several fermented food products and its uses predate our earliest historical records. Its sour and pungent taste is largely attributed to the accumulation of acetic acid. Substrates used for the production of vinegar are usually malt, wart, grape, apple juice or cane juice etc., and the inoculants are the species of *Acetobacter* (*A. orleansis, A. schutzenbachi, A. aceti*). These bacteria are aerobic in nature and are usually found in the wild state growing as pellicles at the liquid interfaces of alcoholic liquor.

Commercial production of vinegar varies considerably in its processing in different countries. In *Orleans process*, commonly used in France, the acetic acid bacteria are allowed to grow on the surface of the wine in partly filled casks. When the wine is slowly converted to vinegar part is drawn off at the base of the cask, the original volume is re-established by replinishing with new wine. Though it is a slow process, it produces vinegar of highest quality.

The most widely used method of vinegar production is *Fring's process* or Trickling process (Fig. 8.3). The apparatus consists of a large wooden tower packed with wood shavings. From the top the alcoholic liquor is distributed over the packing which provides support for the growth of acetic acid bacteria. The liquor is acetified as it trickles down the tower to a reservoir at the base from which it is recycled until the process is complete. Acetification being exothermic, the heat so produced causes the tower to act as a chimney drawing cold air in the base and discharging warm at the top. Air regulator is attached to the base to make the system cool. The method though time-consuming, is simple in operation.

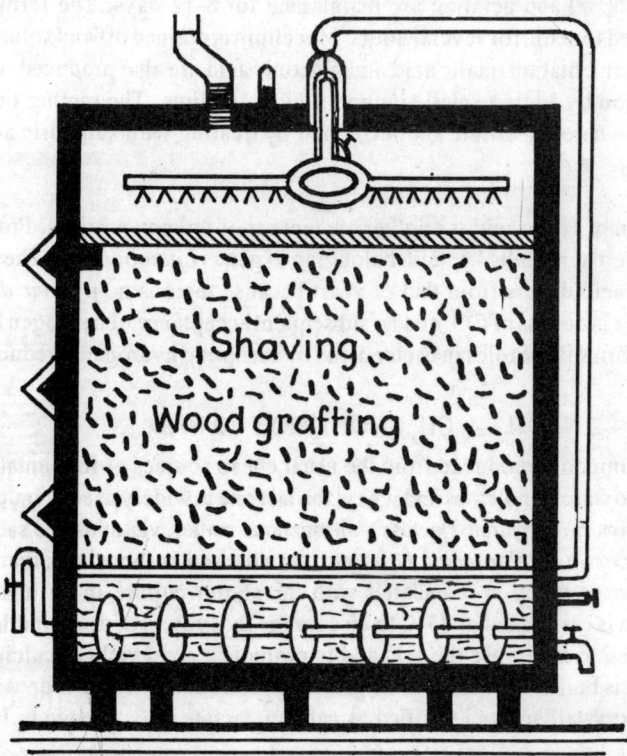

Fig. 8.3   Trickling process or *Fring's process* for
the production of vinegar.

In recent years, vinegar is being produced in *acetator* by submerged fermentation. In this process constant and intense aeration is maintained to keep the organism in a rapid phase of acetification and provision is made for efficient dissipation of heat. The submerged acetification is done both in batch and continuous flow fermentation techniques. Such system is good for producing large quantities of vinegar in a relatively short time and in a minimum of vat space.

Vinegar contains at least 4% acetic acid and small amounts of alcohol, glycerol, ester, reducing sugars, pentosans, salts and other substances. It is used in pickling, in preserving meats and vegetables and in the manufacture of salad dressing. It is a god muscle-toner and is used by ladies and beauticians in European countries.

Citric acid

Amongst many organic acids produced by micro-organisms in industry by partial oxidation of glucose or other substrates, citric acid is the most extensively used one. It is used in soft drinks and other food and medicinal preparations. It is a superior sequestering agent and is employed in the manufacture of ink, dyeing, electroplating and leather tanning.

For the production of citric acid beet molasses containing 48-52% sugar is usually used as substrate. A pH range of 5.5 - 6.5 is adjusted by adding dilute sulphuric acid. Potassium, phosphorus, nitrogen in the form of acid or salt are mixed for the proper growth of the fungus *Aspergillus niger* or *A. wentii*. The ferment medium is allowed to flow by gravity in shallow alumunium pans arranged in trays. Temperature (28-32°C), relative humidity (40-60%) and aeration are maintained for 8-12 days. The fermented liquor is run into settling tank and allowed to settle for several hours. Mycelium is drained off and solution is heated to 80-90°C. During the course of fermentation oxalic acid and gluconic acid are also produced which are precipitated by preferential precipitation by adding small amount of hydrated lime. The mother liquor contains citric acid which is precipitated as calcium citrate and recovered by treating with sulphuric acid.

Lactic acid

Microbial fermentation of lactic acid is another ancient art of unknown origin. Production of lactic acid is largely achieved by microaerophilic bacteria belonging to genus *Lactobacillus*. The bacterial pathway of the fermentation of lactic acid differs from that of yeast because there is no *pyruvic decarboxylase* present in bacteria and hence there is no loss of CO_2 and no subsequent acceptance of hydrogen by acetaldehyde. Instead pyruvic acid itself performs these roles in bacteria and by accepting hydrogen is reduced to single end product, lactic acid.

$$CH_3.CO.COOH + 2\ NADH \longrightarrow CH_3.CHOH.COOH + 2\ NAD^+$$

Lactic acid is commonly produced from the usual cheap sources of fermentable carbohydrate such as hydrolysed corn, potato starch, molasses and whey; the last one is widely used in industry for the manufacture of lactic acid that contains carbohydrate (lactose), nitrogenous matter, vitamins and salts. The homofermentable lactobacilli viz; *Lactobacillus delbrueckii, L. bulgaricus* are used as inoculants that are maintained in skim-milk medium. Pasteurised whey is inoculated with the starter culture in an incubation tank (5-10% by volume). Fermentation is carried out at 43°C to discourage the growth of undesirable organisms. Slurry lime (Ca (OH)$_2$) is also added to neutralise the acid and to promote a good yield of calcium lactate. At the end of fermentation, material is boiled to coagulate the protein (lactalbumin) which is decanted off and filtered. The filtrate is evaporated, crystallised and purified as calcium lactate or converted to lactic acid.

Lactic acid is an odourless, colourless liquid used as an acidulant in confectionary, fruit juices and essences. It is also used in curing of meat, fish products and canned vegetables. Besides, its application in drinking industry, in the form of calcium lactate, plastic and leather industries are also very common.

Pharmaceutical products

Since the introduction of penicillin there has been a growing demand of antibiotics and roughly over 100,000 tons of antibiotics are produced each year in the world. Contribution of fermentation technology for the production of therapeutic compounds is immense. Several antibiotics, vitamins and medically important steroid hormones are now being produced commercially by microbial activities. Recombinant DNA technology has been of immense help in the production of vaccine by isolating and manupulating the genes that direct the synthesis of surface antigens of virulent pathogens and transferring them to the harmless bacteria. Some important antibiotics produced by microbes are listed in Table. 8.4.

Table 8.4
MICROBIAL ORIGIN OF SOME IMPORTANT ANTIBIOTICS

Group	Micro-organisms	Antibiotics
Fungi	Penicillium notatum	
	P. chrysogenum	Penicillin
	P. griseofulvum	Griseofulvin
	Aspergillus fumigatus	Fumagillin
	Cephalosporium sp.	Cephalosporin C
	Paecilomyces varioti	Variotine
Actinomyces	Streptomyces griseus	Streptomycin
	S. nodosus	Amphotericin B
	S. venezuelae	Chloromycetin (Chloramphenicol)
	S. orchidaceous	Cycloserin
	S. rimosus	Oxytetracyclin (Terramycin)
	S. aureofaciens	Chlorotetracyclin
	S. erythrens	Erythromycin
	Nocardia sp.	Ristocetin Vancomycin
	Micromonospora sp.	Gentamycin
Bacteria	Bacillus brevis	Gramicidin Tyrothricin
	B. colistinus	Colistin
	B. licheniformis	Bacitracin
	B. polymyxa	Polymyxin

Many steroid hormones are manufactured by microbial bioconversion. These have obvious advantages over the chemical synthesis. For example, microbial synthesis of cortison, an anti-inflammatory agent commonly used in the treatment of rheumatoid arthritis, reduces the number of chemical reactions from 37 needed for chemical synthesis to only 11 which has reduced the cost of cortison about 400 fold. Some important steroid transformations of commercial importance achieved through microbes are depicted in figure 8.4.

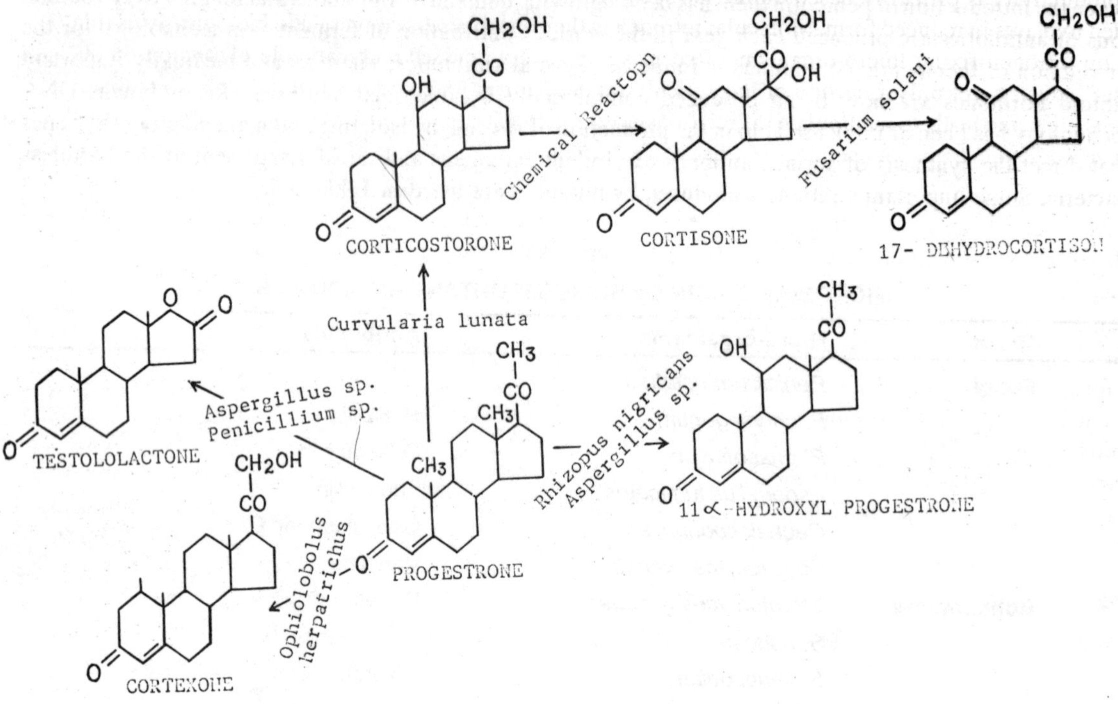

Fig. 8.4 Transformation of steroids by micro-organisms.

Microbial production is the only source of vitamin B_{12}. Many bacteria viz; *Bacillus megaterium, Propionibacterium freudenreichii, Streptomyces olivaceous* are known to synthesise vitamin B_{12} in different phases of fermentation. Besides riboflavin and ascorbic acid are also manufactured to any significant extent microbiologically. However, none of the fat soluble vitamin is produced industrially by microorganisms except *B*-carotené which is converted by animals to vitamin A.

Genetic manipulations of bacteria and yeasts have yielded many useful end products that was not achieved earlier. Human growth hormones (e.g., somatostatin), interferon, insulin, pure viral antigens for vaccines are some of the important microbial products obtained through genetic know-how and engineering. Human insulin produced by microbes is the preferred choice over the commercial insulin obtained from its traditional source - cows and pigs - because of least allergic reaction induced in the recipient host.

MICROBES IN AGRICULTURE

Soil is the largest repository of microorganisms. Many pathogenic and beneficial microbes coexist in the various strata of the mother earth. These contribute significantly to crop productivity. Detailed account of soil microbiology is presented in chapter-IX. In the following account we are presenting the major contribution of microbes for agricultural economy.

Biofertiliser

Crop productivity is primarily based on the nitrogen economy of the soil. This essential nutrient must be provided to plants in reduced form. Molecular nitrogen in the atmosphere is converted to biologically usable forms by nitrogen-fixing micro-organisms. Recent years have witnessed a remarkable expansion of the fertiliser industry, yet only a portion of the agricultural need for nitrogen comes from chemical fertiliser. Attempts are being made to manipulate the relationship between plants and nitrogen fixing micro-organisms to reduce our dependence on chemical fertilisers. The problems has been solved partially by nature itself.

Leguminous plants in their root nodules contain population of *Rhizobium,* a nitrogen fixing bacterium. These plants can grow well in nitrogen poor soil and their cultivation supplements nitrogen requirement for the subsequent non-leguminous crops. Modern approaches seek to enhance the nitrogen fixing ability of the legumes by manoevouring the nodules. Many strains of *Rhizobium* have been developed with superior nitrogen fixing capacity. Introduction of these strains in the soil of different eco-regions is a challenging scientific endeavour. Besides, remarkable success has been achieved by introducing *Azotobacter,* a free living nitrogen fixing bacterium to stick tightly to the root of corn, a non-leguminous plant thereby freeing corn of its requirement for nitrogen fertiliser. The most sophisticated approach of modern agricultural technology is to transfer genes for nitrogen fixation from bacteria to plants. The free living bacterium, *Klebsiella pneumoniae,* that contains all the 17 nitrogen fixation (**nif**) genes in a single large plasmid is extensively studied. The plasmid has been cloned and transferred to yeast *Saccharomyces cerevisiae* and possibilities are being explored to transfer the genes from prokaryote to eukaryote system.

Siderophores

Some strains of *Pseudomonas putida* and *Ps. fluorescens* when inoculated into soil can promote growth of plants. These bacteria get attached to the root hairs and secrete extracellular substances called **siderophore.** These compounds bind iron tightly from the soil and suppress the growth of other micro-organisms near the root depriving them of many essential nutrients. By doing so they reduce the colonisation of harmful bacteria and fungi and help in higher yields.

Mycorrhizae

These are the fungi colonizing the fine root hairs of plants and functionally increase the area of interface between plant roots and soil. During the last decade mycorrhizal studies have revealed that these 'fungus-roots' particularly **Vesicular Arbuscular Endomycorrhizae** (VAM) directly improve the growth of plants especially in the soils of low fertility. Mycorrhiza formation facilitates more efficient uptake of water and nutrients increasing the plant vigor, growth rates and yield. Such ingenious means of circumventing the use of expensive fertiliser are of tremendous economic significance that may help convert the marginal land into agriculturally productive field.

Biological Control

Introduction of living organisms (other than man) in the ecological niches to inhibit or eliminate the harmful effect of pathogenic organisms is the basic principle of biological control. Application of chemical insecticides or pesticides has some obvious disadvantages. Besides being too costly, many of these chemicals have long term toxicity, pollute the environment and accumulate in the food chain manifesting many serious consequences. Their lethal effect is also fairly nonspecific killing beneficial insects and predators. In recent decades several insect specific pathogens have been screened for use as microbial pesticide. For example *Bacillus thuringiensis* is used as an excellent pesticide that produces toxic crystals which are quite effective to control gypsy moth larvae, cabbage worms or tent caterpillars. Likewise, *Baculovirus* (see p.71) is recommended for the elimination of many pathogenic insect population. Many saprophytic bacteria have been successfully tried for the biological control of a number of bacterial diseases like crown gall, bacterial blight etc. Some fungi viz; *Trichoderma viridae, Aspergillus niger, Memnoniella echinata* etc. have been found effective in controlling many fungal diseases in plants.

MICROBES AS POLLUTION INDICATOR

Microbes are highly sensitive to **environment** in which they grow. Their equilibrium is governed by the abiotic component of the environment. A slight fluctuation in the component is readily responded to by the microbes. This responding ability of the microbes is exploited as a tool for the detection of pollutional load in the system. Unlike chemical and physical tests, microbial assays possess many qualities that are useful in pollution study. These are inexpensive, highly sensitive, reproducible and easy to perform. The basic principle of 'microbial assay' rests on the fact that micro-organisms require the same essentials for growth (carbon, nitrogen, energy and micronutrients) as do all other forms of life. Organisms deficient in a specific enzymatic step or pathway that is needed for the biosynthesis of a particular growth factor will grow only when that factor is present in its environment. Such growth limiting factor, if present, in an impure substance (pollutants) is easily recognised by the micro-organisms manifesting their growth pattern. Based on this philosophy polluted zones are identified by detecting the particular micro-organism in the area that serves as biological indicator. This can be well illustrated in the following examples.

Water Pollution

Aquatic systems are prone to three types of pollutions i.e., physical, chemical and biological. Of these, biological pollution is most harmful to us. Physical pollution leads to eutrophication that is threatening the very existence of the freshwater ecosystems. Introduction of inorganic and organic wastes to water bodies in forms like laundry detergents, radioactive wastes, pesticides, fertilisers, mining wastes etc. are the major sources of chemical pollution. Biological pollution, on the other hand, develops from the micro-organisms that enter waters from different sources like human waste, food operations, medical facilities etc. Under normal conditions the water body is able to handle biological materials because heterotrophs digest the organic matter to soluble phosphate, nitrates or sulphate that are used by plants. Problem arises when water becomes stagnant or gets overloaded with waste. Water is then unable to handle the biological materials and becomes polluted. A critical factor operating under such a situation is the increased **biological oxygen demand** (BOD). It refers to the oxygen requirement by the metabolising organisms. As the number of organisms increases, the demand for oxygen increases significantly. Depletion of O_2 in water may lead to death of fish and other aquatic biota. As biological indictor many of the micro-organisms exhibit several morphological, behavioural and physiological changes which can be used as indices for feeling the pulse of the system. For example, actinomycetes, autotrophic bacteria, cellulose digesters, *Bacillus, Paramecium, Euglena* etc. are the normal inhabitants of unpolluted water whereas coliform bacteria, fecal streptococci, blue-green algae, protozoan cysts, *Clostridium* sp., commonly occur in polluted zone. Presence of fecal coliforms, especially *Escherichia coli* in water indicates fecal contamination and possible presence of water borne pathogens.

Air Pollution

Detection of air pollution is largely accomplished by microbes. Recently a bioluminescent bacterium *Photobacterium phosphoreum* has been successfully employed to detect the level of the pollution in air. Cells of the bacterium are treated with pollutants generated by irradiation of a gas mixture of cis-2-butane and nitrogen oxide. At frequent intervals the gas mixture is replaced with nitrogen. The resulting anaerobic condition reduces the bacterial luciferin and the enzyme reduction produces a flash of light, the intensity of which is measured with great sensitivity. As the available oxygen is removed the light intensity also decays logrithmically. The result is compared in terms of luminescence on passing polluted air over the bacterial cells. In cigarette industry the toxicity is checked by passing the cigarette smoke over the protozoan colonies of *Paramecium aurelia*. The survival time of the colonies is measured and regulated by filter as per requirement. This system is a convenient model for testing materials designed to reduce the toxicity of cigarette smoke.

MICROBES IN CANCER RESEARCH

Bacteria have been recruited to 'sniff out' substances that are possible carcinogens (cancer causers). About 90 per cent of the known carcinogens are also mutagens. A potentially carcinogenic substance could therefore, be detected by determining its ability to cause mutations. Bacteria provide an exceptionally sensitive, rapid and inexpensive system for detecting the mutagenic properties of such substances. Working on these lines, Bruce Ames at California University, USA developed a procedure to determine the potentiality of a chemical to induce mutation in bacteria and thus being a cause of cancer. The test is popularly known as *Salmonella*/microsome Ames test. The procedure involves the inoculation of standardised number of histidine-requiring strains *(auxotrophs)* of *Salmonella typhimurium* into a plate containing the culture medium lacking histidine. Normally the *Salmonella* strain will not grow as genes for histidine synthesis are lacking. When histidine is supplemented into the medium the bacterial colony starts developing till histidine is completely exhausted. A mutation in the correct gene could reverse the original mutation restoring the test bacteria's ability to synthesise the needed nutrients and in the course develop the **revertant colonies.** A significant increase in the number of revertants indicates the mutagenic potentiality of the chemical. If the mutagen is soaked in the centrally placed filter disc of the plate, a characteristic accumulation of mutant colonies appears around the filter disc. This test has been quite reliable in detecting hundred of compounds of mutagenic activity within a short time like hair dyes, cigarette smoke, pesticide, mycotoxins etc.

MICROBES IN BIODEGRADATION OF RECALCITRANT

Modern technology has produced many such materials which are not readily degraded by the natural biogeochemical cycle. Many low molecular weight compounds such as solvents, refrigerants, propellants, flame retardants, pesticides, herbicides etc. are continuously pouring into the nature and are hard to be metabolised by microbes. These chemicals are known as **recalcitrants.** Plastics are one of such synthetic compounds that have accumulated substantially in the biosphere in recent years. Microbes have not yet developed the ability to decompose them. Many of the non-degradable compounds are injurious to humans and their hazardous consequences have compelled the manufactureres to curtail their production (e.g., DDT). Realising the potentialities of the microbial decomposition of these recalcitrants the scientists are devoting serious attention to develop the capabilities of the microbes to accomplish this task. One approach is a continuous culture enrichment technique that allows the microbe to grow in a culture supplemented with the molecularly allied degradable compounds and gradually substituting it by the target substance. Another approach is to develop genetically engineered bacteria that can covert the nondegradable pollutants to useful products. Microbial degradation of petroleum product is a recent scientific challenge that has been taken up on priority basis especially in view of the recent Gulf war of 1991 in which a huge amount of petroleum was spilled in the sea causing a serious threat to the environment. Recombinant DNA technology has produced plasmid-carrying strains of marine bacteria that rapidly metabolise crude oil even in cold ocean.

MICROBES AS FUEL RESOURCE

The present crisis of petroleum and other non-renewable energy sources have led the scientists all over the world to search for alternative means to supplement the deficit. Microbes may provide a partial solution to this problem by converting sewage and other wastes into usable fuel. Harvesting of methane gas as a byproduct of sewage treatment to generate electricity is a new scientific endeavour. Biotransformatin of sunlight into usable fuel is another promising possibility. Photosynthetic micro-organisms generate hydrogen gas, a clean fuel that produces only one waste product, water, on burning. Laboratory experimentations have yielded encouraging results. Theoretically use of microbial enzymes to generate electricity is an exploring area of research. The electrons released by microbial oxidation is allowed to concentrate on one side of a membrane where an external circuit is provided to flow these electrons to produce electricity.

Cellulose reserve is an ideal biological renewable resource for energy. Ethanol and methane are the two fuels that can be collected as byproducts of biodegradation of huge cellulose waste. Tapping these resources would have triple advantages-energy production, waste disposal and single cell protein production.

MICROBES IN BIOMINING

Approximately two-third of our oil reserves are not recoverable. Microbes are considered as ideal tool for promoting oil released from the porus rocks. Microbes are thought to help in oil extraction in the following ways : (i) acids produced by bacteria dissolve calcareous cementing materials and decompose sulphate minerals directly resulting in greater porosity of oil-bearing sands, (ii) bacteria produce CO_2, CH_4 and H_2 which are dissolved in soil reducing the viscosity and increasing the gas pressure (iii) they may reduce detergent that help to release oil. These theoretical possibilities are, however, yet to be worked out. In modern copper and uranium mines metals from low grade ore are extracted with the aid of bacterium *Thiobacillus*. This bacterium is a chemolithotrophs (rock-eaters). This causes break down of sulfide minerals to sulphate (SO_4^{++}) and causes the conversion of copper in the ore to the cupric state (Cu^{++}). These ions spontaneously react and yield soluble copper sulfate. The solution is then processed to harvest the dissolved copper in its pure elemental form.

MICROBES IN SPACE RESEARCH

Space microbiology or **exobiology** deals with the possible occurrence of micro-organisms in outer space and on planets. Microbes are used for food and energy and also for maintenance of suitable oxygen-carbondioxide balance in space vehicle. To study the effect of weightlessness, gravitational force, extreme temperature and other extra-terrestrial conditions on life, microbes are used as ideal experimental tools.

SUGGESTED READINGS

Alcamo, I.E. (1987). Fundamentals of Microbiology, 2nd ed. Banjamin/Cummings Publishing Co. Inc., California.

Delanney, A. and H. Erni (eds.) (1965). The World of Microbes, Vol.4., Encyclopedia of the Life Sciences. Doubleday, Garden City, New York.

Larry, McKane & Judy Kandel (1986). Microbiology - Essentials and Applications. Mc.Graw Hill International Book

Riviere, J. (1978). Industrial Applications of Microbiology. Halstead Press, New York.

Wilkinson, J.F. (1986). Introduction to Microbiology (Basic Microbiology Series, Vol.I) 3rd ed. Blackwell, Oxford.

9

MICROBES IN SOIL

Soil is the loose, weathered, consolidated covering of the earth's crust that is formed from the disintegration and decomposition of rocks by physical, chemical and biological processes. The soil particles range in size from 0.002 mm (of clay) to > 2.00 mm (of gravel) and are characterised by their water holding capacity, chemical composition and microbial spectrum.

Soil constitutes a complex environment comprising all the three phases- solid, liquid and gas - arranging in different proportions constituting different soil types. The soil atmosphere differs from that of air as it contains more of carbondioxide and lesser oxygen. Under certain conditions when gaseous exchange is disrrupted by permanent water-logging, not only carbondioxide but methane, hydrogen and

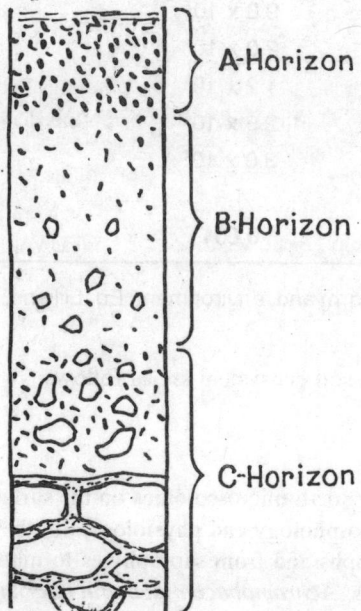

Fig. 9.1   A soil profile.
A - Horizon - litter, decaying organic matter and mineral material
B - Horizon - Mineral material with additions from upper horizon
C - Horizon - Broken fragments of parent rock.

hydrogen sulfide also accumulate. All these have profound effect on the microbial world present in the vicinity of the soil. The activities and compositions of microbes are dependent on the soil profile (F'g.9.1) that contains varying proportions of organic matter in different horizons. As a matter of fact, the soil acts as a buffer, reducing the range of physical conditions as compared with those found on and above its surface. Its effectiveness as a buffer depends on its chemical composition and physical make-up. The clay, in this regard, are particularly important because of their ability to absorb nutrients, enzymes and profuse population of micro-organisms. The more fertile soils, on the whole, ensure a more constant supply of water and nutrients than do the infertile ones.

SOIL POPULATION

The dynamism of soil is largely determined by microbial components which in association with plant parts make this non-living system more than an inert geological deposi:. It has been estimated that one gram of agricultural soil contains millions of bacteria, thousands of fungal propagules and hundreds of protozoa and algae. In (Table-9.1) an estimate of the numbers of main categories of micro-organisms is presented with an estimate of the biomass of each of the groups. In general, bacteria though predominate numerically in an agricultural soil, biomass of fungi is larger than that of non-filamentous bacteria.

Table 9.1

NUMBER AND BIOMASS OF SOME ORGANISMS IN THE TOP 15 cm. OF AGRICULTURAL SOIL

Group	Number of Organisms per gram	Biomass g/m^2
BACTERIA	9.8×10^7	160
ACTINOMYCETES	2.0×10^6	160
FUNGI	1.2×10^5	200
ALGAE	2.5×10^4	32
PROTOZOA	3.0×10^4	38
NEMATODE	1.5	12
EARTHWORMS	0.001	80

(Source : Micro-organismes - function, form and environment. Ed. Lillian E. Hawker and Alan H. Linton 2nd ed. Edward Arnold Publ. Ltd. 1979).

The important groups of microbes in a soil ecosystem are as follows:

BACTERIA

Bacteria tend to grow as individuals or small micro-colonies on the surface of soil particles. They exhibit almost the whole range of bactetrial morphology and physiologically they range from aerobes to obligate anaerobes, from hetrotrophs to autotrophs and from saprophytes to mutualistic and parasitic symbionts. Members of the genera *Pseudomonas, Achromobacter* and *Bacillus* are mostly found in aerobic soil whereas *Clostridium* is specifically restricted to the anaerobic condition. One of the most important activities that bacteria perform in soil is that of nitrogen fixation. These bacteria are *Clostridium, Azotobacter, Rhizobium, Nitrosomonas, Nitrosococcus, Nitrobacter* etc. Some of the sulphur bacteria eg., *Thiobacillus* oxidise sulphur to sulphate which are utilised by plants or removed from soil by rain water.

The bacteria that constitute majority of the soil population and are least affected by soil amendment are *Arthrobacter* and *Agrobacterium*. The size of the bacterial population depends not only on the nutrients available but on other environmental factors as well. Temperature, moisture, pH, gas contents of the soils - all induce variations in bacterial population. Besides, the spatial distribution of bacteria is also governed by the depth of the soil which is a reflection of varying organic matter content in different horizons.

ACTINOMYCETES

Most actinomycetes are soil inhabitants. Numerically these are only second to those of bacteria that go on increasing with the warmth of the climate and decrease with the depth of the soil. Like most other bacteria their growth is largely favoured by alkaline condition and improved soil aeration. In acid and water-logged soils they are relatively scarce. Many of the actinomycetes can grow in a soil which is not suitable for the growth of other organisms. They usually flourish after fast growing micro-organisms have transformed organic residues into a dark homogenous mass, rich in lignin, hemicellulose and proteins. Actinomycetes compete effectively with other micro-organisms owing to their ability to degrade the residual nutrients available in the soil.

FUNGI

A large number of fungi belonging to diverse taxonomic groups are the natural inhabitants of soil. In general *Mucor, Mortierella* and *Rhizopus* of Mucorales and *Penicillium, Aspergillus, Cladosporium, Fusarium, Alternaria, Trichoderma, Cephalosporium, Verticillium* etc. of deuteromycetes occur commonly in soil. Some dark coloured hyphae, possibly sterile mycelia of Dematiaceae are also frequently encountered in different soils. The biomass of fungi in cultivated soil often exceeds that of its counterparts and in acidic environment they are often numerically dominant too.

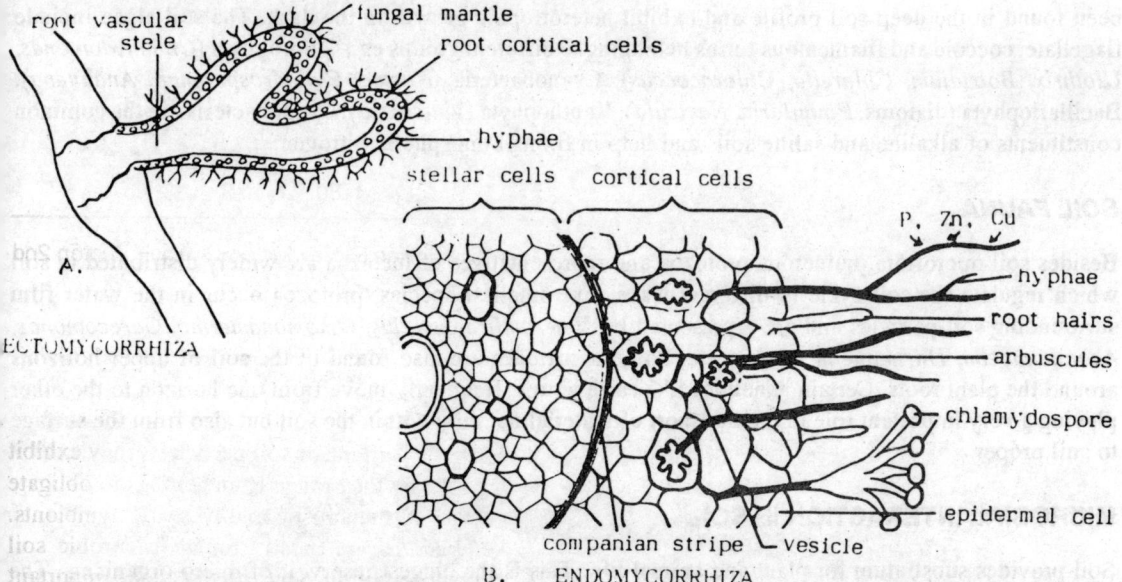

Fig. 9.2   Association of ectomycorrhiza and vesicular arbuscular mycorrhiza with plant roots.

A. Ectomycorrhiza produces short branched root-lets covered with fungal mantle; hyphae do not penetrate the cells.

B. Endomycorrhiza penetrates inside cells and forms arbuscles and vesicles.

The major activity of soil fungi is to decompose the complex organic matter into the simpler ones for the formation of manure in the soil. Depending on their preference of substrates for utilisation, these are classified as **sugar fungi** (mainly the members of phycomycetes), **cellulose fungi** (phycomycetes and fungi imperfecti), **lignin decomposing fungi** (basidiomycetes), humus fungi, **coprophilous fungi** (dung fungi), predaceous fungi etc.

Some of the fungi associate themselves with the roots of higher plants in the form of **mycorrhiza** (fungus root). These mycorrhizae may be **ectotrophic** (growing as an external cover around the roots) or **endotrophic** (penetrating into the host cell) and contribute beneficially to the plant growth(Fig. 9.2).

Like bacteria the fungal activity in soil is largely affected by water level. Fungi are comparatively resistant to dessication and are therefore major share-holder of the soil ecosystem of arid regions. Water-logging in saturated soil and consequent reduction of oxygen concentration inhibit most of the filamentous fungi because majority of the fungi are strict aerobes. Fungi occur most frequently in the upper soil horizon where there is more organic matter and like many other organisms show relative fall in number with the depth of the soil. The abilities of fungal mycelium to ramify in the soil particles and bind these together improve the texture of clay soil and promote the soil fertility in a large way.

The presence of yeast has also been established in most of the soils. The important genera of soil-yeast are *Candida, Cryptococcus, Torula, Torulospora, Pichia* etc.

ALGAE

The motile algae, specially flagellate and some of the members of Volvocales are the common inhabitants of even temporarily wet soil or in rain puddes that last for a few days. Most of the algae are usually confined to the top soil surface. However, there are reports that they can occur underneath the stones and even within the cracks of the stones. Many members of bacillariophyta and chlorophyta have been found in the deep soil profile and exhibit heterotrophic growth in the dark. The soil algae include flagellate, coccoid and filamentous forms belonging to different groups eg., Chlorophyta (*Chlamydomonas, Ulothrix, Botridium, Chlorella, Chlorococcus*), Cyanobacteria (*Nostoc, Cylindrospermum, Anabaena*), Bacillariophyta (diatoms, *Pinnularia, Navicula*), Xanthophyta (*Vaucheria*). Cyanobacteria are the common constituents of alkaline and saline soils and help in fixing atmospheric nitrogen.

SOIL FAUNA

Besides soil microflora, numerous protozoa and representatives of metazoa are widely distributed in soil which regulate the soil cycle in different ways. The smallest species, protozoa occur in the water film surrounding soil particles and are represented by *Heteromitra cucullus, Oikomonastermo, Carecomonas, Acanthamoeba, Hartmanella hyalina* etc. Some nematodes are also found in the soil of upper horizons around the plant roots. Certain annelides like earthworms frequently move from one horizon to the other playing a very important role in the transport of material not only within the soil but also from the surface to soil proper.

MICROBIAL INTERACTION IN SOIL

Soil provides substratum for plant and animal life. This is the biggest reservoir of micro-organisms. The varied soil micro-organisms interact with one another in a number of ways. Symbiosis', commensalism, mutualism, antagonisms etc. are some of the common terms used to describe the various types of interactions among the micro-organisms. However, most of the interactions between the soil organisms are

For terms see the glossary

competitive in which the better adapted partner comes out as a real beneficiary of the competition. As a result of interaction, many micro-organisms may flourish as predominant forms while the poor competitors may be eliminated completely. Thus a sort of microbial succession occurs which alters the soil profile considerably paving ways for other microflora to colonize.

In a soil ecosystem besides the mutual interactions among microbes, a distinct interaction effect is also ensued between the plant and the microbes. The zone of increased microbial activity immediately around the root region is known as **rhizosphere**, a term initially introduced by Hiltner (1904). The concept of **rhizosphere effect** on plant growth has received considerable significance in the field of agriculture. Jenny and Grossenbacher in 1962 presented an electronmicrograph of the young root surface that revealed the covering of mucilage layer having intense microbial concentration. In a demonstrable experiment, Parkinson and his associates (1965) reported that when an inert nylon thread of the same dimension as that of plant roots was burried in soil alongwith the dead roots for sometime, micro-organisms accumulated around the dead roots and not the nylon threads. This shows that the micro-organisms have a definite affiliation to the organic matter of plant residues.

Plant roots provide conditions for the maximum microbial activity and it is this region through which soil organisms exert their influence on the plant as well. Consequently the plant also exerts a sort of **"return influence"** on microflora of the rhizosphere affecting its colonization in the vicinity. Proximity of the soil to the root, depth of the soil profile, type of plant species, habit of plants, age, maturation and fruiting all are the major determining factors for rhizosphere microflora. In this plant-microbe interaction root excretions manifest a pronounced effect on the microbial spectrum. Root liberates some water soluble compounds which stimulate the germination of spores of many fungi. Quite often they liberate certain antimicrobial agents which may also play an important role in microbial antagonism. Generally the influence of the root extends for only a few millimetre in the soil. Its effect on the microbial population is expressed as the **R/S ratio** which is conventionally defined as the number of organisms in rhizosphere soil as compared to the number in the same soil beyond the influence of the roots. Table-9.2 shows a comparative picture of microbial distribution in the rhizosphere and non-rhizosphere soil of wheat field. The R/S ratio of bacteria in a cultivable soil is usually much higher indicating pronounced bacterial activity in the soil.

Table 9.2

NUMBER OF MICRO-OGANISMS IN RHIZOSPHERE OF WHEAT

MICRO-ORGANISMS	RHIZOSPHERE	NON-RHIZOSPHERE	R/S RATIO
	NO./g DRY WEIGHT		
BACTERIA	1200×10^7	5×10^7	240.0
ACTINOMYCETES	46×10^6	7×10^6	6.6
FUNGI	12×10^5	1×10^5	12.0
PROTOZOA	2.4×10^3	1×10^3	2.4
ALGAE	5×10^3	27×10^3	0.2

(SOURCE : Rouatt, J.J., Katznelson, H and Pyne, TMB (1960) Proc. Soil Sci. Soc. Amer. **24**, 271-273)

The influence of plant on the rhizosphere flora is mainly due to sloughing-off of dead cells from the growing root tips and older parts of the root. The exudation of organic compounds from the young roots in the form of amino acid, organic acids and soluble sugars provides easily assimiable nutrients for the microflora which help in their colonization around the root zone. In return, micro-organisms influence the plant roots in a number of ways. Root branching and root hair production are the general effects of rhizosphere flora. Besides, the physiology of root is also affected by the production of growth stimulating substances (eg., Indole-acetic acid) as well as by the secretion of polypeptide membrane-active antibiotics which stimulate further leakage of plant cell contents into the soil. The production of carbon dioxide in the rhizosphere, formation of organic and inorganic acid and removal of oxygen from the environment affecting root respiration and ion-uptake are the major influences of microflora on the plant system. Above all, nodulation of *Rhizobium* by legume roots for the fixation of atmospheric nitrogen and improving phosphorus supply to the plants are the well realised contribution of rhizosphere fungi. Nevertheless, the significance of rhizosphere micro-organisms for the plant has been a matter of speculation. Their detrimental effects in releasing toxic substances and inducing plant diseases are important from pathological angles. However, there is substantial evidence to suggest that plants grown in sterile soil do not grow that well as those in soil inoculated with micro-organisms. Also, such plants are more susceptible to infection by reintroduced soil-borne plant pathogens. Investigations on these lines may unravel many intricacies of soil-microbe-plant interactions.

BIOGEOCHEMICAL RECYCLING

The earth is a closed biosphere. No new matter in significant amount enters the sphere. Only light energy comes to the earth in a continuous stream and is released back into the space. This light energy regulates all biological phenomena on the earth. Living systems use this energy to continue life through growth and reproduction. Since no new atoms are being added to the earth, living systems must use the available atoms again and again. This repeated utilisation of energy is known as *biogeochemical recycling* in which inorganic molecules are combined to form the organic compounds of living oganisms. If there were no ways to recycle this organic matter back into its inorganic form the organic material would have piled up as a dead load on the earth. Here decomposers play a vital role in running the cycle. If the microbes are prevented from doing their job, organic materials would accumulate in the form of deposit of oil, natural gas, and coal.

The soil population is largely responsible for the transformations of elements which are essential for the fertility of the soil and which ensure the removal of natural litter from the surface of the earth. Carbon, oxygen, nitrogen, sulphur, phosphorus etc. are some common elements found in all the living organisms and these must be recycled after the death of the organisms. Soil micro-organisms act upon these constituents and after their mineralisation make the elements available for reuse. In the protoplasm of living organisms these elements are generally found in a reduced state. On mineralisation they come in oxidised state and function as a source of energy for oxidation and as electron acceptor in oxidation reactions. Most of the transformations of elements are cyclic and comprise many sequential steps. Some stages result in an amelioration of the soil whereas others decrease the fertility of soil. In some instances, stages in a cycle are mediated only by an individual organism, others take place as a result of the combined action of wide varieties of micro-organisms. Some important transformations of elements are enumerated below:

CARBON CYCLE

Carbon is the most important element of the biological world. About half of the dry weight of living organism is composed of carbon. This huge amount of carbon comes from atmospheric carbon dioxide. The atmospheric CO_2 is converted to organic carbon mainly by the action of photoautotrophic organisms like higher green plants on land and algae in water, though some heterotrophs also use some small quantities of this gas. Some amount of carbon dioxide derived from dissolution of carbonates and biocarbonates is utilised by some chemolithotrophs. Carbon dioxide utilised by these organisms is subsequently transformed into various cell components like carbohydrate, proteins, fats, nucleic acid etc. Once the atmospheric carbon dioxide is fixed, it becomes non-available for the generation of new plant-life. Therefore, it is essential for carbonaceous materials to be decomposed and returned back to the atmosphere otherwise the little stock of carbon dioxide in atmosphere (0.3%) would be exhausted. The return of carbon is brought about mainly through the decomposition of organic carbon in nature; a small amount comes back by way of respiration. A simplified representation of the cycle is given in fig.9.3.

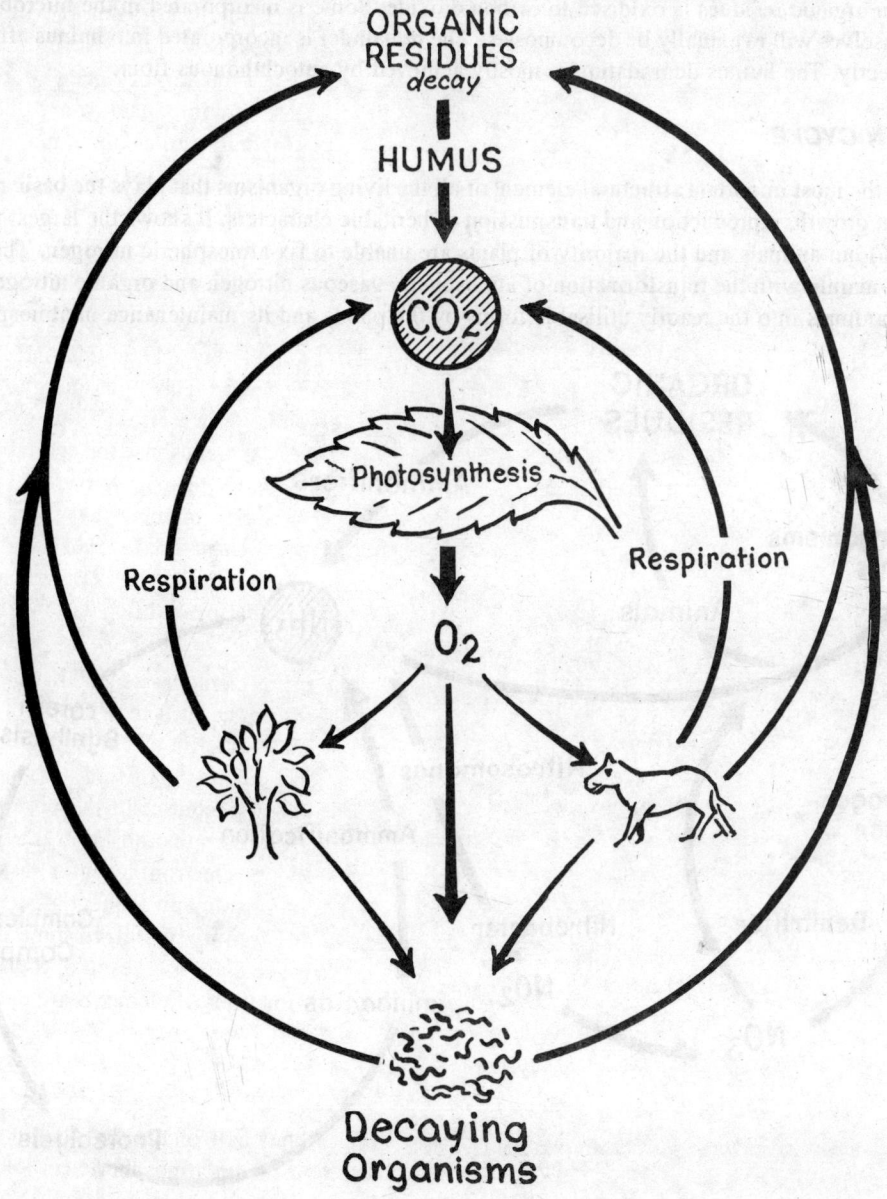

ORGANIC
RESIDUES
decay

HUMUS

CO_2

Photosynthesis

Respiration

Respiration

O_2

Decaying
Organisms

Fig. 9.3 The carbon cycle

The major forms of carbon compounds added to the soil are plant and animal remains in which the carbon is incorporated in high molecular weight compounds. In plants the compounds of this type are cellulose, hemicellulose, lignin, fats, waxes, proteins and nucleic acids. Microbial attack on these compounds primarily depends on the specific chemical nature of the residue, environmental conditions and the nature of the underlying soil. Of the environmental conditions moisture and temperature are particularly important. The underlying soil determines to a large extent the population available for degradation and their ability to liberate specific enzymes for the degradation of the high molecular weight polymer. Most of the carbon contained in organic residues is oxidised to carbon dioxide. Some is incorporated in the microbial tissues which themselves will eventually be decomposed. The remainder is incorporated into humus affecting the fertility directly. The humus degradation is mostly achieved by autochthonous flora.

NITROGEN CYCLE

Nitrogen is the most important structural element of all the living organisms that plays the basic role in cell metabolism, growth, reproduction and transmission of heritable characters. It shows the largest volume of air (78.08%) but animals and the majority of plants are unable to fix atmospheric nitrogen. The nitrogen cycle deals mainly with the transformation of atmospheric gaseous nitrogen and organic nitrogen of dead plants and animals into the readily utilisable forms by the plants and its maintenance in atmosphere. The

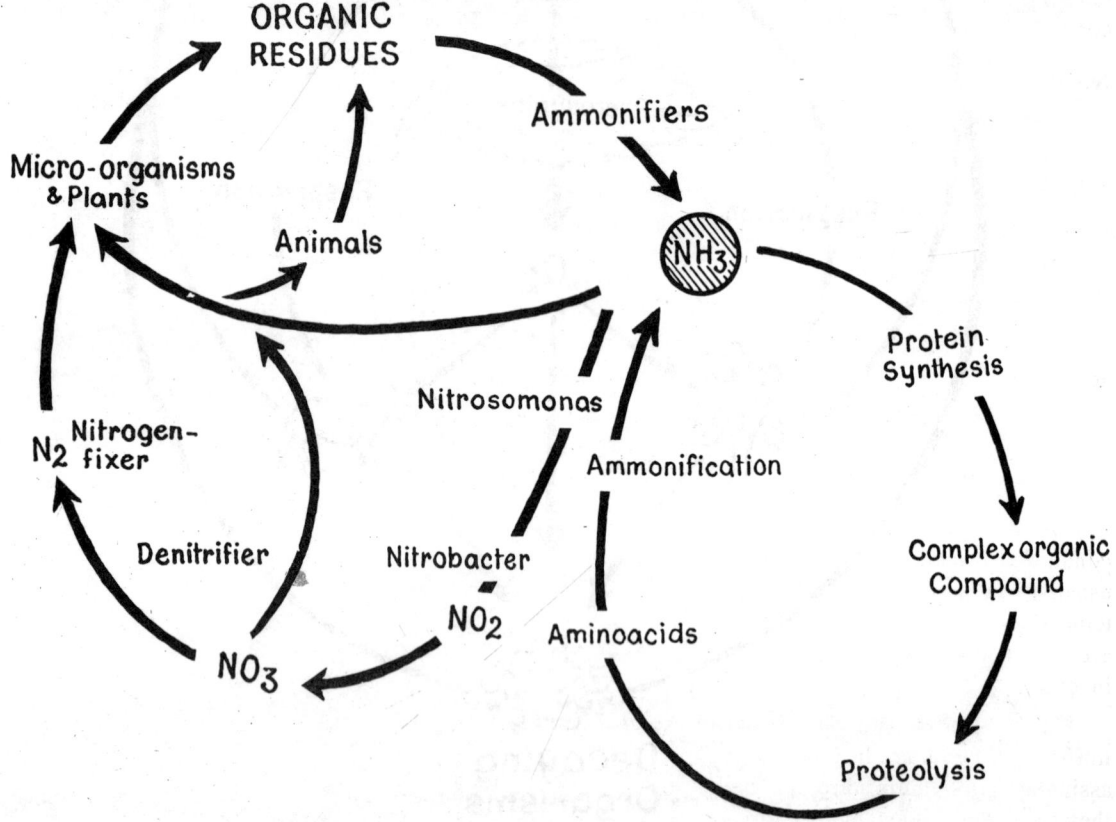

Fig. 9.4 The nitrogen cycle

activities of many micro-organisms make nitrogen available to plants in an assimiable form. These organisms act in an analogous way to those involved in the carbon cycle which make carbon dioxide available. However, there are two differences between the carbon and nitrogen cycle in terrestrial environment. Firstly, carbon is made available to plants in gaseous form as carbon dioxide, whereas nitrogen is utilised by plants in the form of ammonium or nitrogen ions. Secondly, most of the micro-organisms involved in the carbon cycle are heterotrophic whereas many of those taking part in the nitorgen cycle are autotrophic. The cycling of nitrogen is presented in fiqure.9.4 which involves five major processes viz., ammonification, nitrification, nitrate reduction, denitrification and nitrogen fixation. The details of various processes are as follows:

Ammonification

Release of ammonia from organic nitrogenous compounds is termed as ammonification. Proteins and other organic nitrogenous compounds of living and dead organisms are decomposed by soil micro-organisms with the help of various proteolytic enzymes to amino acids. The amino groups ($-NH_2$) are split-off to form ammonia. Under aerobic condition the amino acids are deaminated (oxidative deamination) to liberate ammonia whereas under anaerobic condition amino acids are converted to amines and related compounds (putrefaction) which are finally oxidised in the presence of air with liberation of ammonia. Many fungi, actinomycetes, clostridia and some species of *Pseudomonas, Bacillus, Proteus* are known to accomplish the ammonification more efficiently out of which *Clostridium* spp. are the major contributors of anaerobic condition.

Nitrification

The oxidation of ammonia to nitrate is called nitrification. The process consists of two steps i.e., from ammonia to nitrites and then to nitrate. The oxidation is achieved chiefly by organisms of only two strictly autotrophic genera : *Nitrosomonas* and *Nitrobacter*.

Nitrosomonas is a small Gram-negative rod with-polar flagella and is strictly aerobic. It mediates the overall reaction:

$$2NH_4^+ + 3O_2 \longrightarrow 2NO_2^- + 4H^+ + 2H_2O$$

Besides *Nitrosomonas*, species of *Nitrosococcus, Nitrosospira* and *Nitrosocystis* are also known to accomplish the transformation of ammonia to nitrite. The intermediates in this oxidation are not yet confirmed, but one is probably **hydroxylamine**.

Nitrobacter is slightly smaller than *Nitrosomonas* and converts nitrites to nitrate-

$$2NO_2^- + O_2 \longrightarrow 2NO_3$$

Formation of nitrate from ammonia is a very sensitive reaction fully dependent on the environmental condition particularly being carried in neutral and alkaline soil. At soil pH value below 6 ammonium ions usually accumulate and in very alkaline soil high concentrations of nitrate may be found. Extremes of temperature and drought may also cause accumulation of nitrite. Since both *Nitrosomonas* and *Nitrobacter* are strict aerobes, water-logging of the soil and concomitant reduction in gaseous exchange render them inactive.

In most of the cultivable soils nitrification is the rule but in may orchards, forests and grasslands the nitrifying bacteria are inactive because of soil or climatic condition. In such cases, plants must therefore assimilate ammonium nitrogen. Unlike nitrate and nitrite, the ammonium ions have the advantage that they are chemically stable in acidic soil and they are usually bound to negatively charged clay particles and are not easily leached out of the soil. Only a small amount of the ammonium ions bound to clay particles are exchangeable and are available for assimilation. Its concentration beyond a certain level may be toxic to plants. Hence, from agricultural point of view this ammonium assimilation is not desirable.

Nitrate reduction

The nitrification process is reversed by many micro-organisms which are capable of reducing nitrate to nitrite and then to ammonia. The process is called **nitrate reduction** and involves several reactions. The overall result is -

$$HNO_3 + 4H_2 \longrightarrow NH_3 + 3H_2O$$

This process is also known as *assimilatory nitrate reduction* because in this process cellular nitrogen is obtained through ammonium assimilation. Many species of bacteria, yeast and fungi are known to assimilate nitrate nitrogen through this process.

Denitrification

Many micro-organisms are capable of reducing nitrates to nitrites and subsequently to gaseous nitrogen (eg., molecular nitrogen or nitrous oxide) causing a reduction in the level of soil nitrogen. This is called denitrification. The process differs from the earlier nitrate reduction because it does not include the assimilation of ammonium ions by plants or micro-organisms or loss by leaching, rather in this case nitrate competes as an alternative electron acceptor to oxygen especially under anaerobic condition or in an abundant supply of organic matter.

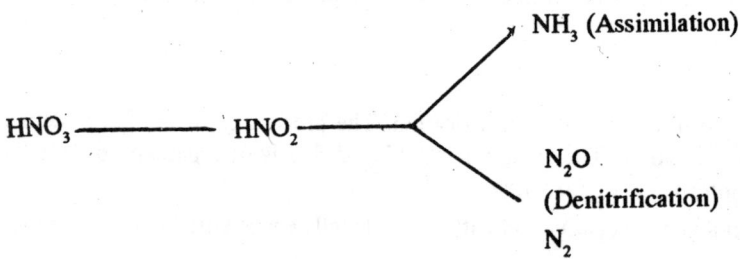

When nitrate is used as a source of electron acceptor there is a net loss of nitrogen from the soil. The process is therefore called *dissimilatory nitrate reduction*.

$$2NO_3^- + 10 \, e^- + 12 \, H^+ \longrightarrow N_2 + 6H_2O$$

Improved aeration usually reduces denitrification. Seasonal flooding of land or over-irrigation of poorly drained land leads to nitrogen loss by denitrification. Some nitrate reducing anaerobic micro-organims contribute significantly in denitrification. The species of *Pseudomonas, Achromobacter, Clostridium, Serratia, Micrococcus* and *Thiobacillus* are important in this respect. Fungi and actinomycetes are probably not involved though they may be more important in acidic forest soil.

Nitrogen Fixation

The combination of nitrogen and hydrogen to form ammonia through biological means is known as **nitrogen fixation**. Prokaryotes are the only organisms that are able to tap the nitrogen reservoir of the atmosphere and fix molecular nitrogen. Either as free living organisms or in symbiosis with higher plants they can carry out reactions that result in the incorporation of nitrogen into organic compounds. According to rough estimates, the symbiotic nitrogen fixation of leguminous plants gains a nitrogen yield of 100-300 Kg N/hectare/annum and the free living cyanobacteria contribute about 30-50 Kg N/ha/annum. In addition, significant amount of bound nitrogen may reach the soil via precipitation from the atmosphere that amounts to 3-30 Kg N/ha/annum.

Nitrogen fixing micro-organisms

There are two main groups of nitrogen fixing organisms- (1) symbiotic nitrogen fixers -those capable of fixing molecular nitrogen by living in the roots of leguminous plants and (2) non-symbiotic nitrogen fixers - those capable of fixing molecular nitrogen independently of other living organisms. The classical free living micro-organisms capable of fixing nitrogen are species of *Azotobacter* (*A. chroococcum; A. beijerinckii, A. vinelandii, A. macrocytogenes, A. agalis*)and *Clostridium* (*C. pasteurianum, C. butyricum, C. aceticum, C. felsineum, C. beijerinckii*). Besides, many blue-green algae, chemoautotrophic and photosynthetic bacteria are also known to fix atmospheric nitrogen independently (Table-9.3).

Table - 9.3

NON-SYMBIOTIC NITROGEN FIXING MICRO-ORGANISMS

Heterotrophic bacteria	Chemoautotrophic bacteria	Photosynthetic bacteria	Blue-green algae
Azotobacter	Methanobacillus	Chlorobium	Anabaena
Aerobacter	Desulfovibrio	Chromatium	Anabaenopsis
Achromobacter		Rhodomicrobium	Calothrix
Pseudomonas		Rhodopseudomonas	Cylindro-
Clostridium		Rhodospirillum	spermum
Bacillus polymyxa			Tolypthrix

All the species of *Azotobacter* are Gram-negative aerobic bacteria that can fix atmospheric nitrogen in alkaline soil more effectively. They are, however, unable to fix nitrogen below pH 6. One aerobic bacterium *Beijerinckia* has been recently identified to be capable of fixing nitrogen at pH as low as 3. Clostridia are anaerobic and are able to fix atmospheric nitrogen in a wider range of pH (5.0-9.0). Nitrogen fixation by free-living cyanobacteria is of considerable importance at least in rice fields and other flooded soils.

Among the various symbiotic association of micro-organisms to higher plants *Rhizobium* legume association is the most common one for nitrogen fixation. Rhizobia occur as free-living Gram-negative rods in soil and are strictly aerobic. They are grouped in the following six major species on the basis of their host specificities and other characteristics (Table-9.4).

Table - 9.4

RHIZOBIUM - LEGUME ASSOCIATION

Rhizobium leguminosarum	Pea, lentil, vetch
R. meliloti	Alfalfa , sweet clover
R. trifoli	Red and white clover
R. phaseoli	Bean
R. lupini	Lupines
R. japonicum	Soybean

Rhizobium of cow-pea group and lotus group show their specificities towards cowpea and lotus respectively.

NITROGEN FIXATION BY SYMBIOTIC BACTERIA

Fixation of atmospheric nitrogen by symbiotic bacteria primarily depends on the formation of root **nodules.** The relationship between the nodule formation and nitrogen assimilation was first demonstrated by Hellriegal and Wilfarth in 1888.

Nodulation in leguminous plants

In the development of nodular structure the initial infection of bacteria is restricted to the newly growing

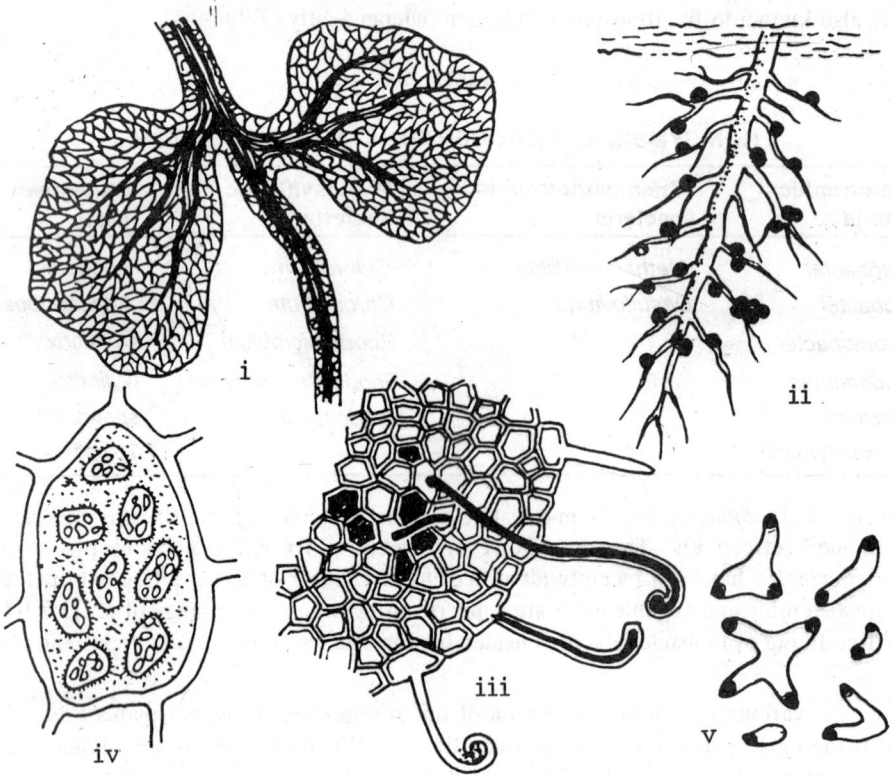

Fig. 9.5   Symbiotic nitrogen fixation in root nodules of leguminous plants.

    i. Section through fully developed nodules

    ii. Root of pea with nodules

    iii. Penetration of bacteria and growth of infection thread

    iv. Section through a cell filled with rhizobia

    v. Various shapes of bacteria inside the cell.

root hairs. This is largely accomplished by the release of plant exudates stimulatory to the infecting bacteria around the root zone (Fig.9.5). The chemical nature of the excretion is unknown but the substances are presumed to be growth factors or energy substrates necessary for the initiation of infection. Leguminous plants contain **lectins.** These are glycoproteins that bind specific **polysaccharide.** It is assumed that the lectins are localised on the outer layer of the root hairs. The outer layer of the rhizobial cell wall, on the other hand, contains species-specific polysaccharides. A recent hypothesis, under investigation, suggests that the compatibility of the partners is achieved by the interaction between the lectins on the surface of the root hairs and the surface polysaccharide of the *Rhizobium*. The enzyme polygalacturonase is also believed to play its role in root hair infection, possibly by acting on the pectin of the plant cell wall to permit bacterial penetration. However, the products released by the roots are highly selective for the stimulation of bacteria of the specific cross-inoculation group.

In most legumes bacteria enter at or close to the growing point of the root hairs in the form of infection thread which undergoes deformation or curling under the influence of some microbial products, presumably **indoleacetic acid**. The narrow infection tube is typically surrounded by a wall of cellulose synthesised by the host. The infection thread branches into the central portion of the developing nodules and the bacteria ultimately are released into their symbiont's cytoplasm therein to multiply. Following the release a period of rapid cell division takes place in the host cell. The final structure consists of a central core containing the rhizobia and a surrounding cortical area in which is found the plant vascular system.

The plant cells in the central portion of the nodule possess twice the chromosome number characteristics of the host cells. The disomatic tissue probably originates from the disomtic cells of the uninfected roots which upon the approach of the invading bacteria are stimulated to multiply forming tetraploid cells. However, the exact mechanism for such transformation is still not very clear.

Once liberated from the infection thread into the root cytoplasm the rhizobia assume a peculiar morphology, a cellular form that has been termed as **bacteroid**. They are swollen and irregular appearing as star, clubbed or branched shaped structures. Recently *in vitro* studies have been carried out to induce artificial bacteroid formations to assimilate nitrogen in the absence of the plants on the presumption that nitrogen fixation and bacteroid formation are the two essential components of symbiotic nitrogen fixation. The findings, however, wait unequivocality. Concomittant with the formation of bacteroids a pink colouration develops in the cytoplasm of the host cells due to the formation of haemoglobin related pigment-**leghaemoglobin**. Only those nodules that contain leghaemoglobin are able to fix molecular nitrogen in its ferrous form. Its formation is a specific result of symbiosis; the prosthetic group, the **protohaem** is synthesised by the bacteroids and the protein component is formed by the plant cells. Leghaemoglobin has high affinity for oxygen and is assumed to facilitate the diffusion of oxygen ensuring sufficient oxygen supply for bacterial energy demand, which would otherwise inhibit nitrogen fixation.

Nodulation in non-leguminous plants

Many dicotyledonous plants that are not members of the Leguminosae are also known to contain root nodules with the ability to fix molecular nitrogen. These nodules also exhibit symbiotic partnership with prokaryotes, especially actinomycetes and member of the genus *Frankia*. Some common non-leguminous nitrogen fixing plants are *Casurina equisetifolia, Alnus, Hippophoe* and *Ceanothus. Myrica gale, Elaeagnus, Shepherdia, Coriaria* are comparatively less efficient nitrogen fixers. In nitrogen poor habitats these plants contribute significantly in enriching the soil; a few of these amount to 150-300 kg N/ha/annum. The root nodules of these woody plants quite often attain the size of a tennis ball. In *Casurina* the nodules consist of a loose bundle of thickened rootlets whose growth is negatively geotrophic. Like leguminous plants infection of root occurs from the soil via root hairs. However, only the outer parenchyma cells of these plants are infected by the symbionts. Other symbiotic manifestations are almost similar to that of leguminous plants.

Symbiotic nitrogen fixation with Cyanobacteria

Besides being a major contributor in nitrogen fixation as independent unit, many cyanobacteria also occur as nitrogen-fixing partners in symbiosis with higher plants. In the water fern *Azolla,* cyanobacteria are contained in the tissue space of leaves. The symbiotic partner is *Anabaena azolae*. In comparison to the free living *Anabaena* that contains only 5% heterocyst, the symbiotic *Anabaena* trichomes have 15-20% heterocysts that is indicative of the effective nitrogen fixation. The nitrogen yield of symbiosis of *Azolla* and *Anabaena* is approximately 300 Kg N/ha/annum. A similar symbiosis occurs between liverworts (*Blasia pusilla, Anthoceros punctatus, Peltigera* sp.) and *Nostoc*.

MECHANISM OF NITROGEN FIXATION

The mechanism of nitrogen fixation is a highly debated issue. Our concept about the mechanism is chiefly based on the laboratory experiments with non-symbiotic nitrogen fixing organisms, especially the cell free preparation of *Clostridium pasteurianum*. The biochemistry is, however, similar in all the organisms that have been studied so far.

Nitrogen fixation is a reductive process in which ammonia is the first demonstrable product. The process is mediated by an enzyme complex called nitrogenase. This consists of two sub-units, a **molybdenum-iron-sulphur protein** and an **iron-sulphur protein**. The enzyme and the process of nitrogen fixation are extremely sensitive to oxygen and thus there must be a mechanism that can protect the nitrogenase from high partial pressure of oxygen. Reducing power and energy both are necessary for nitrogen fixation. These are supplied by photosynthesis, respiration or fermentation. The energy is supplied in the form of ATP and the reducing power in the form of reduced pyridine nucleotide or ferredoxins via flavodoxin-containing carriers. A general scheme for nitrogen fixation is presented in fig.9.6. Nitrogenase can reduce not only molecular nitrogen ($N\equiv N$) but also the other substrates like acetylene azide, nitrous oxide, cyanide, nitrite, isonitrite and protons. Acetylene reduction provides the simplest method for demonstrating nitrogenase activity. All nitrogen fixing organisms are able to reduce acetylene to ethylene which can be quantified chromatographically.

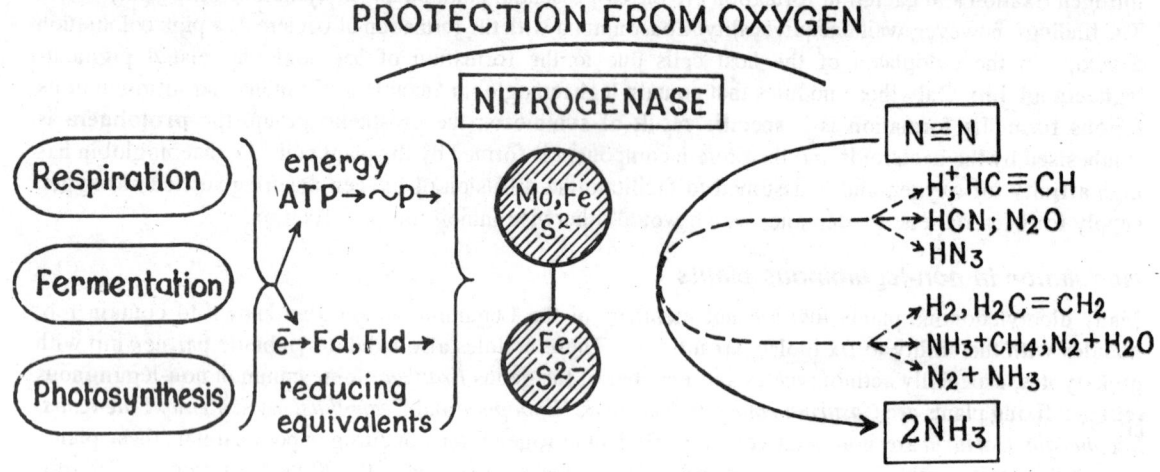

Fig. 9.6　Schematic representation of nitrogen fixtion.

Fd - ferredoxin; Fld - flavodoxin.

Nitrogenase has the property of an ATP-dependent H_2-evolving hydrogenase. In the absence of molecular nitrogen it catalyses the reduction of protons to molecular hydrogen. Even in the presence of nitrogen, part of the reducing equivalents is utilised for proton reduction and hydrogen production and the rest for the reduction of nitrogen.

$$8[H] + N_2 + 2H^+ \longrightarrow 2NH_4^+ + H_2$$

In addition to nitrogenase almost all nitrogen-fixing bacteria contain a hydrogenase which can activate the hydrogen relesed from the nitrogen reaction. The hydrogenase thus serves to rechannel the reducing equivalents into metabolic reactions and conserves them for ATP generation.

SULPHUR CYCLE

Sulphur is one of the most abundant elements in nature that is found in free as well as in combined states. In the combined form it occurs in both inorganic and organic combinations. In agricultural soils sulphur meets the requirement of both the soil population and crop plants. The main supply of sulphur to the environment is derived from the parent rocks, sulphur springs, organic wastes, volcanoes and industrial gases. Accumulation of sulphur in soil or water bodies has considerable detrimental effects on the flora and fauna. Cycling of sulphur is therefore necessary to maintain the equilibrium of soil elements for the benefit of the crop plants. Micro-organisms are more varsatile in utilising most of the sulphur compounds and are also instrumental in making sulphur available to higher plants in utilisable form. A brief outline of sulphur cycle is presented in fiqure 9.7.

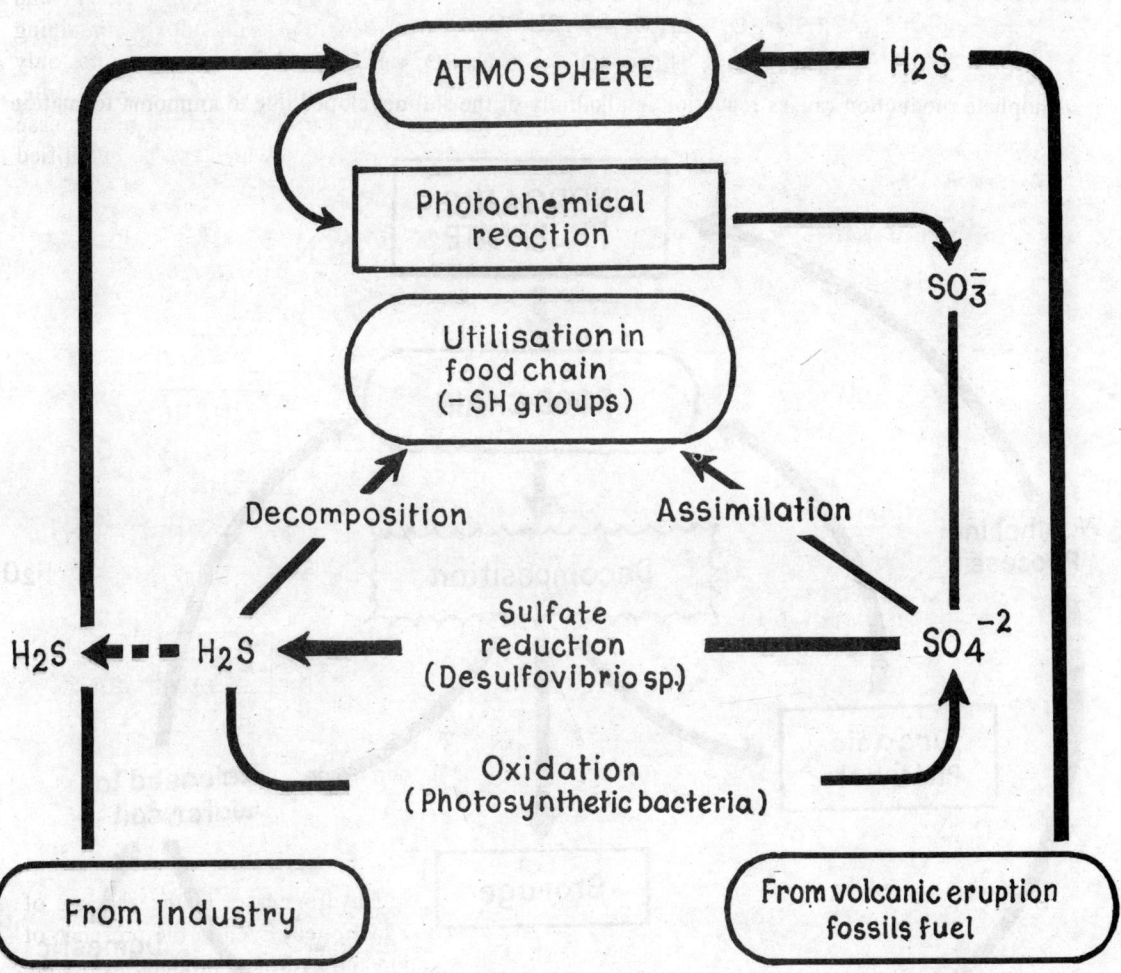

Fig. 9.7 The sulphur cycle.

Oxidation of sulphur compounds

Sulphur may exist in many oxidation states (eg., hydrogen sulphide) and many micro-organisms, particularly photosynthetic and chemolithotrophic bacteria can carry out the conversion. Under anaerobic condition photosynthetic sulphur bacteria belonging to family Chlorobiaceae and Chromatiacae cleave H_2S to elemental sulphur

$$CO_2 + 2H_2S \xrightarrow{\text{-L-i-g-h-t-}} (CH_2O) + 2\,S + H_2O$$

Hydrogen sulphide acts as an electron donor for CO_2 reduction in photosynthesis.

Under aerobic condition chemolithotrophic bacteria of family Thiobacteriaceae (eg., *Thiobacillus theooxidans*) can oxidise sulphides, elemental sulphur, sulphite etc. to sulphuric acid and sulphate which is beneficial to plants in many ways.

$$2\,S + 3O_2 + 2H_2O \longrightarrow 2H_2SO_4$$
$$5\,Na_2S_2O_3 + H_2O + 4O_2 \longrightarrow 5\,Na_2SO_4 + H_2SO_4 + 4\,S$$

Sulphate production causes reduction in alkalinity of the soil developed due to ammonia formation

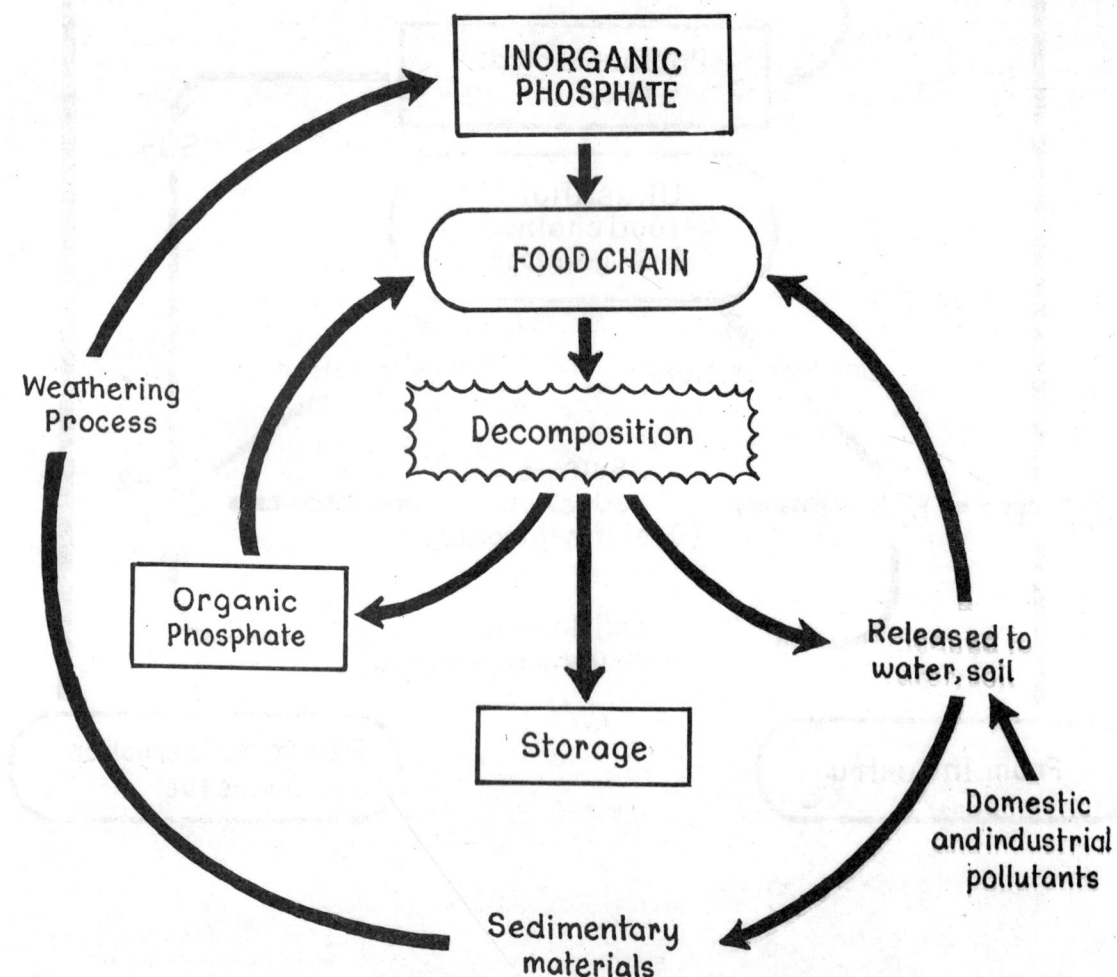

Fig. 9.8 The phosphorus cycle.

by micro-organisms. Some filamentous bacteria such as *Beggiatoa* and *Thiothrix* may live on the surface of muds rich in hydrogen sulphide and they oxidise it to elemental sulphur.

Reduction in sulphate

The organisms chiefly responsible for reducing sulphate in soil are members of the genus *Desulfovibrio* which use sulphate rather than oxygen as a terminal electron acceptor. In agricultural soils these organisms do not play a very important role owing to aerobic condition but in water-logged paddy field soils where anaerobic condition prevail, they are responsible for the generation of hydrogen sulphide which may damage the roots of rice plant. However, these sulphate reducing organisms are economically important in a number of ways. They are responsible for causing corrosion to underground iron pipe by the removal of hydrogen which normally forms a protective layer around these pipe in the absence of air. They also play a part in liberating oil from oil-bearing states.

PHOSPHORUS CYCLE

Phosphorus is present in organic forms in the protoplasm of living organisms. It is needed in substantial quantity for nucleic acid synthesis and for phopholipids and high energy phosphorus compound like ATP. Soils are usually deficient in this element and its supply is largely supplemented by plant and animal residues or by phosphate fertilizers. On the death of living organisms the organic phosphorus is changed to inorganic phosphoric acid which is further converted to insoluble aluminium.

Organic residues added to the soil are degraded by the heterotrophic microflora. The phosphorus content present in the residue determines whether phosphate is made available to crop plants or it gets incorporated into microbial tissues. Phosphate is made available to plants by microbes which liberate organic acids to dissolve insoluble inorganic phosphate compound in the soil. Mycorrhiza association of plants plays an important role in the uptake of phosphorus. Scheme of phosphorus cycle is presented in figure 9.8.

SUGGESTED READINGS

Alexander, M. (1977). Introduction to soil microbiology, 2nd Edn. John Wiley, New York & London.

Brock, T.D. (1966). Principle of Microbial Ecology. Printice-Hall, Englewood Cliff, New Jersey.

Burges, A. and Raw, F. eds (1967). Soil Biology. Academic Press, London.

Fenchl, T. and Blackburn, J.H. (1979). Bacteria and mineral cycling. Academic Press, London.

Gray, T.R.G. and Williams, S.T. (1971). Soil-micro-organisms. Oliver and Boyd, Edinberg.

Postgate, J.R. (1982). The Fundamental of nitrogen fixation. Cambridge University Press, London.

Sprent, J.I. (1979). The biology of nitrogen-fixing organisms. McGraw Hill, London.

10

MICROBES AND DISEASES

Men, microbes and diseases are the three integral components of human history. Man's initial experience of the microbes was perhaps through the diseases. With the establishment of germ theory of diseases a new branch of science, **Medical Microbiology**, came into being that deals with the causes, symptoms, etiology and control of diseases.

A '**disease**' is any process or condition that disturbs the normal functioning of an organism. This is a broad definition. A broken arm, a diabetic patient, a person suffering from goitre (endocrine disorder) or scurvy (deficiency of vitamin C) are all diseased conditions but these do not necessarily involve micro-organisms. In the present chapter we would confine our discussion to the diseases that are essentially caused by the micro-organisms.

HOST - PARASITE RELATIONSHIP

Human body harbours millions of micro-organisms. Some organisms establish permanent relationship with the body (e.g., *E. coli* in large intestine) while others are transient (e.g., Streptococci) and still others are of pathogenic nature. Under normal conditions a direct competitive relationship exists between the human body and these parasites. If the host exerts its supremacy due to increased host-resistance, it remains healthy and the parasite is either driven away or assumes a benign relationship with the host. On the contrary, if the host loses the competition then due to increased pathogenicity, the disease develops. Thus diseases are the outcome of competition between the host and the parasite. However, in many instances the competition is latent or subclinical. Such sub-clinical condition is ideal for the parasite because it drains nutrients from the host indefinitely. Such an organism is called **benign parasite**. When pathogen overcomes the body's resistance barrier and succeeds in establishing a clinical (diseased) condition it is known as **inept parasite**. Sometimes a potential pathogen becomes part of normal flora of the host body and gets associated with a particular organ. When unusual circumstances develop within the body, the micro-organisms manifest disease symptoms in the host. Such agent is known as **opportunistic parasite**. *Diplococcus pneumoniae* is a normal inhabitant of the pharyngeal region of throat. Wet and cold condition facilitates the pathogen to produce the disease. *Candida albicans*, a common yeast is present in the female urogenital system. Application of antibiotics suppresses the bacterial population of the body but the yeast, being resistant to antibiotics, becomes aggressive and produces **candidiasis**. Thus the normal flora performs a valuable function for the body in that it tends to keep the number of potential pathogens at a low level. In most of the **cases**, the causative agent for the disease is a single pathogen. Sometimes two or more pathogens interact to produce an infectious condition in the body. Such synergistic effect is very much evident in dental carries (tooth-decay) where streptococci, lactobacilli, acid producing bacteria and proteolytic organisms exert their cumulative effect to dissolve the tooth enamel and penetrate the fissures causing destruction of **dentin**. *Entamoeba histolytica*, the causal organism of amoebic dysentry, is effective only in the presence of *Escherichia coli*, a harmless organism of human intestine.

TUG OF WAR

Human body is fully exposed to microbial infections. Micro-orgnisms can invade the body at any stage of life and during the course of interaction a fight between the host and the parasite is invariably ensued. In order to achieve supremacy both the partners formulate strategies to win over the game. Body's defense system tends to contain the aggressiveness of the pathogen by adopting some specific measures. However, under certain circumstances the normal benign organisms may turn pathogenic when body defenses fail or become weak. AIDS is a common example where the defense mechanism of the host completely collapses allowing even the benign organism to play its role.

Pathogenesis

Normally a pathogen manages to infect the host tissue physically.The infectiveness of the micro-organism depends on its ability to get established within the host by overcoming the defense barriers such as antibodies or phagocytes. The skin, the respiratory, digestive and urogenital tract may be the major portal of entry for the pathogen. Contaminated water or food, the biological vector or fomites are the common sources of infection. On infection the microbial cells introduce a variety of chemical substances to the host that become antigenic to that tissue. Virus capsid, bacterial capsule, cell wall and flagella of the bacterium, residual particles of the organism etc. serve as the potent antigens that are spontaneously responded to by the host producing specific antibodies.

Once inside the host body, the pathogen releases many chemical components, metabolic products and enzymes that counteract the normal body defenses. The invasiveness of the micro-organism is frequently associated with the production of extracellular enzymes. These enzymes are referred to as **aggressins** that have the ability to break down cells or connective tissue components. Some of the common aggressins produced by bacteria are listed in table-10.1.

Table 10.1

SOME SPECIFIC ENZYMES PRODUCED BY BACTERIA AND THEIR ROLE IN PATHOGENESIS

Enzymes	Bacteria	Mode of Action
Hyaluronidase	*Streptococcus pyogenes* *Diplococcus pneumonae* *Clostridium perfringens*	Breaks down connective tissue increases permeability of tissue space
Leucocidin	*Streptococcus pyogenes* *Staphylococcus aureus*	Kill leukocytes
Haemolysin	*Staphylococcus aureus* *Streptococcus pyogenes*	Destroys RBC
Lecithinase	*Clostridium perfringens*	Causes lysis of RBC and other tissue cells.
Streptokinase	*Streptococcus pyogenes*	Digests the fibrin of blood clots
Collagenase	*Clostridium perfringens*	Dissolves collagen.

Besides producing specific enzymes, the pathogen also releases certain poisonous substances, the **toxins**, that can destroy the susceptible tissues and produce clinical symptoms. Toxins are cellular components or metabolic products of the organisms that damage or interfere with the activity of tissue cells. Toxins are of two types - **exotoxins** and **endotoxins**. Microbial cell secretes exotoxins into its environment whereas it retains the endotoxins. Thus endotoxins do not exert their effect on the host cells until they are released by the death and disintegration of the microbial cell. Exotoxins are largely produced by Gram-positive bacteria and like typical proteins are sensitive to temperature, alcohol, formaldehyde and dilute acids. These can be subjected to denaturation converting to non-toxic form, the **toxoids**. Toxoids retain most of the chemical features of exotoxins and contain the immunising power without having toxic property. These are therefore used for artificial immunisation. Production of exotoxins results in various clinical disorders in host. *Clostridium botulinum* and certain staphylococci are responsible for food-poisoning (botulism) when swallowed directly. The neuortoxin produced by the pathogens causes paralysis of respiratory muscles. The exotoxin **tetanospasmin** is produced by *Clostridium tetani* that causes muscle spasm and rigidity of jaw (lock-jaw). α-toxin released by *Clostridium perfringens* causes gas-gangrene producing necrosis and haemolysis of the affected part together with distension of the tissues. Diphtheria, cholera, plague etc. are some of the dreaded diseases produced by exotoxins of the invading pathogens.

In contrast to exotoxins, endotoxins are complexes of lipid, polysaccharide and proteins. These are chiefly produced by Gram-negative bacteria by the autolysis of the cell. Endotoxins are resistant to heat, alcohol and dilute acids and do not form toxoids. Most of the endotoxins are antigenic i.e., they elicit the formation of antibodies but the neutralising antibody is far difficult to obtain than that of exotoxin. *Salmonella, Shigella* and other enterobacters are the common endotoxin producers that cause gastro-intestinal disorders coupled with increasing haemorrhage and swelling of tissues. Irrespective of their origin all endotoxins show similar activities in the host and their effects during disease duration are non-specific and varied.

Host's Defense

Eversince the pathogen comes in contact with the host, body's defense mechanism gets activated to ward-off or eliminate the pathogen. Human body responds to foreign intrusions by adopting three dimensional strategies: (i) immediate destruction of pathogen on surface layer (ii) immediate immune response and (iii) long-term immune response. The first one is non-specific that exists in all the human beings and displays the general resistance mechanisms of the body. The second and third are specific and relatively more complex defense lines that develop only in response to the presence of a particular pathogen and are directed solely at that organism. In the present section we would discuss only the general resistance mechanism of human body. The immunological responses would be dealt in succeeding chapter.

Non specific resistance can be broadly categorised in four classes :

Mechanical Defense

The skin and the mucous membrane are the most important mechanical barrier for the pathogens. Unless these are penetrated disease can not establish. The sweat gland of the skin secretes certain body wastes that are noxious to invading micro-organisms. Cells of mucous membrane along the linings of respiratory passage secrete mucous which traps heavy particles and microbes of the air. After filtration, even if the microbes enter the body it is immediately acted upon by the stomach acid which has a pH of about 2.0 and acts as a natural barrier for the pathogens. In the gastrointestinal tract itself, as a result of **peristalsis** (constant rhythmic action of the muscles) majority of the pathogens are flushed out from the body through diarrhea. Though expulsion of parasites by this process is a defense mechanism of the host, very often dehydration associated with the flushing may lead to another complications.

Cellular Defense

This mechanism involves phagocytosis that implies the ingestion of foreign particles. It is the major form of non-specific defense in the body. Shortly after the verification of germ theory of disease, Metchnikoff proposed the theory of phagocytosis which was acclaimed by the world's scientists. This led him to the award of Nobel prize in 1908. The cells involved in the process are called phagocytes. These are the **polymorphonuclear cells** (PMN), monocyte of the circulatory system (see the box 3) and **reticulo endothelial system** (RES) of the body. They include Kupffer cells of the liver and macrophages of the spleen, bone marrow, lymph nodes, brain and connective tissues. Some phagocytic cells are fixed macrophages while others are wandering cells.

Box 3

BLOOD COMPONENTS OF HUMAN CIRCULATORY SYSTEM

I. SERUM	-	minerals, salts, protein and other organic substances in blood fluid.
II. PLASMA	-	serum with clotting agent (fibrinogen, prothrombin)
III. CELLS	-	Number of cells/mm^3

			Percent
Erythrocyte (RBC)	men : 4,500,000 - 5,500,000 women : 4,400,000 - 5,000,000		
Leukocyte	5000 - 9000		
Polymorpho- nuclear cell (granulocyte)	neutrophil	2875 - 5175	55 - 60
	eosinophil	100 - 180	2
	basophil	50 - 90	1
Mononuclear cells (agranulocyte)	monocyte	525 - 600	5 - 8
	lymphocyte	1625 - 2925	30 - 35
Platlets	**2,50,000 - 4,00,000**		

Phagocytic cells consciously seek out a micro-organism that manages to penetrate the epithelia. They interact with the organism just to destroy it. The chemical attraction between the two is however, still obscure. Phagocytosis begins with an invagination and pinching of the cell membrane to form a phagocytic vesicle or **phagosome**. (Fig.10.1). The phagosome fuses with a lysosome, an organelle that contributes digestive enzyme, **lysozyme** and an acidic pH to the digestion process. Lysosomal substances increase the permeability of the capillaries which brings more phagocytes to this area that engulf the foreign materials to complete the process. Sometimes the interaction between the parasite and phagocyte is also enhanced by the presence of antibodies. These protein molecules attach to the parasites and facilitate the adherence to the phagocytic cell at a specific receptor site. In other instances components of the complement system (see p. 154) bind the parasite to the phagocytic cells and perform phagocytosis. Enhanced phagocytosis is called **opsonisation** and the antibody or complement components that encourage it is termed **opsonins**. However, sometimes the micro-organism can grow inside the phagocytic cell. In that case the phagocyte actually protects the micro-organism from chemical action and also helps in the dissemination of the infection throughout the body. For example, sometimes *Mycobacterium tuberculosis* is distributed from the lungs to bone marrow or spleen by this process and after the death of the phagocytes, contributes to the development of bone-tuberculosis.

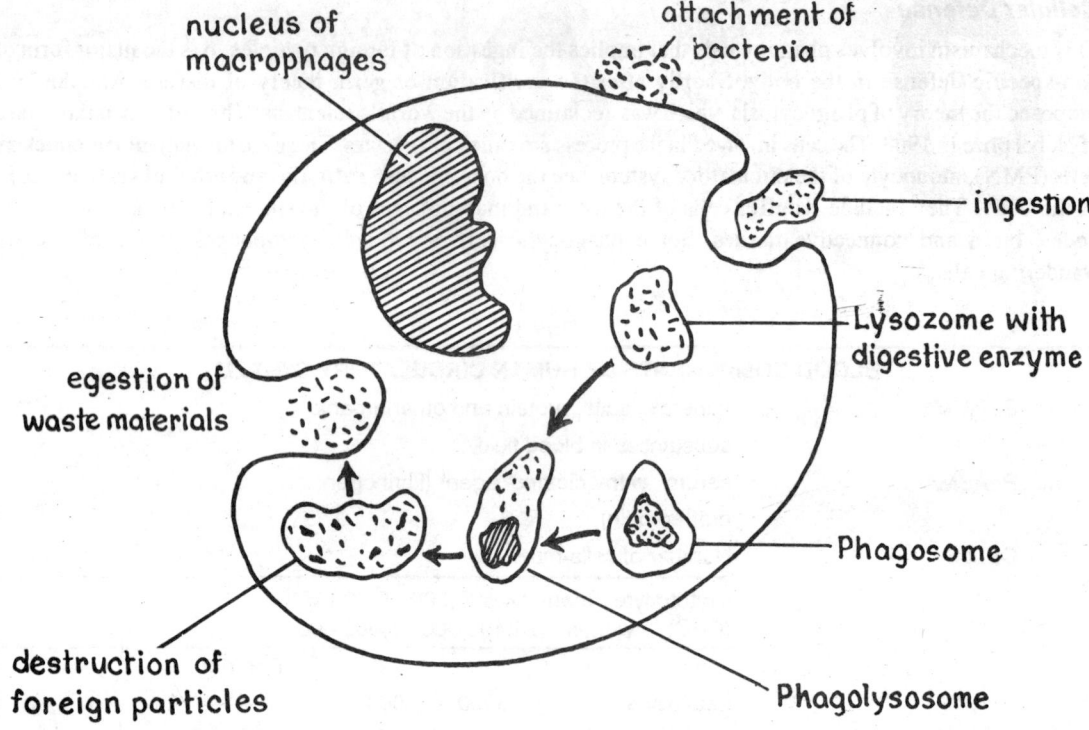

Fig. 10.1 The mechanism of phagocytosis of bacterial cells performed by phagocyte.

Physiological Defense

This is the third line of defense of the human body involving the active participation of various physiological constituents like enzymes, hormones, inorganic salts, polysaccharides, protein etc. One of the common examples of the involvement of enzymes in body's defense is the secretion of lysozyme in tears and saliva (first described by Alexander Fleming who later discovered penicillin). Eyes, being the most exposed part of the body are highly susceptible to infection but the tears (secretion of lacrymal gland) containing enzyme lysozyme is bactericidal in nature. The enzyme acts on the peptidoglycan layer of the bacterial cell wall providing mechanical as well as physiological barrier for the invading microorganisms. The lower gastrointestinal tract harbours a number of Gram-negative and Gram-positive bacteria. Gram-negative bacteria are mostly non-pathogenic and many contribute in the synthesis of vitamin K (an important constituent for blood clotting). Bile salts secreted by the liver, inhibit the concentration of Gram-positve bacteria considerably providing an unusual osmotic condition for the bacterial growth. However, out of all the constituents the most important physiological defense factor is the immunological response that is mediated by protein in the form of **globulin** (antibody) against the antigen. The antigen-antibody response is a complicated phenomenon. The mechanism will be discussed later (see p. 153). In general the antigenic substances on approaching the lymphoid tissues stimulate the later to respond spontaneously causing certain lymphocytes to differentiate into two kinds of cells -i) short-lived plasma cell, which is responsible for the synthesis of antibody globulin and ii) long-lived memory cells which remember how to differentiate plasma cell of a particular antigen.

The neutralisation of antigen by specific antibody is summarised in figure-10.2. Based on the type-reactions antibodies are categorised as :

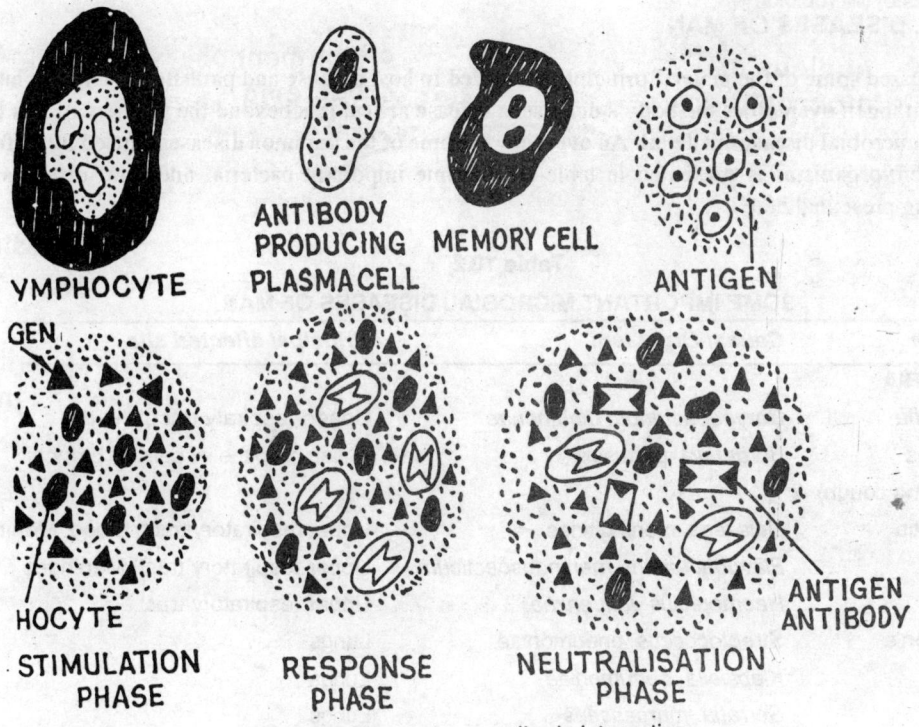

YMPHOCYTE

ANTIBODY
PRODUCING
PLASMA CELL

MEMORY CELL

ANTIGEN

GEN

HOCYTE

STIMULATION
PHASE

RESPONSE
PHASE

NEUTRALISATION
PHASE

ANTIGEN
ANTIBODY

Fig. 10.2 The antigen-antibody interaction.

(i) **agglutinins** - antibodies that immobilise the antigens causing clumping (or agglutination) enhancing phagocytosis.

(ii) **precipitins** - that combine with soluble antigens to convert them to solid particles.

(iii) **opsonins** - that combine with infectious agent to enhance phagocytosis.

(iv) **antitoxins** - that combine with toxin molecules to neutralise the toxicity.

(v) **lysins** - that enhances lysis (bursting) of antigenic cells.

Inflammatory Defense

This is a co-ordinated defense mechanism of the body involving the active participation of all the previous three defense mechanisms. This can be well illustrated by the following example.

Infection of *Staphylococcus aureus*, a pus-forming bacterium, results in a series of sequential events. The pathogen enters the body through the skin abraison penetrating the mechanical barrier of the body and gets established in the blood vessels. The blood vessel becomes dialated and blood plasma and tissue fluid accumulate resulting in the swelling of the zone. A **plasmin** clot may be formed by the conversion of soluble protein **plasminogen** into insoluble plasmin and the area is walled off. Dialation increases the rate of blood flow and the site becomes red. The pathogen releases leukocidin and haemolysin that destroy the host's leukocytes and RBC. These later on serve as the nutrients for the invading organisms; carbohydrate is degraded, organic acids are produced lowering the pH of the area. The host in turn responds to the infection by accumulating the WBC at the spot. Neutorphils adhere to the vessels close to injury and migrate through the wall to begin phagocytosis of the irritants. Subsequently body's immunological system becomes functional and specific antibodies are produced. Opsonins are formed and the process of opsonisation is ensued. As a result of the combat the dead cells (both of host and pathogens) ooze out from the infection site in the form of a mixture of plasma, dead tissues, leukocytes and dead bacteria.

BACTERIAL DISEASES OF MAN

We have discussed some of the general principles involved in host defense and parasite invasion in human body. If the pathogen overcomes the body's defense, a disease arises. It is beyond the purview of this book to deal all the microbial diseases in detail. An overview of some of the common diseases caused by different groups of micro-organisms is presented in table-10.2. Some important bacterial and viral diseases are, however, being presented here.

Table 10.2

SOME IMPORTANT MICROBIAL DISEASES OF MAN

Disease	Causal Organism	Principal affected site
BACTERIA		
Diphtheria	*Corynebacterium diphtheriae*	Upper respiraty tract, Heart
Pertussis (whooping cough)	*Bordetella* pertussis	Ciliated epithelial cells of bronchi
Meningitis	*Neisseria* meningitides.	Upper respiratory tract, Blood, Meninges
	Flavobacterium meningospecticum	Upper respiratory tract, Meninges
	Haemophilus influenzae	Upper respiratory tract
Pneumonia	*Streptococcus* pneumoniae	Lungs
	Klebsiella penumoniae	Lungs
	Serratia marcescens	Lungs
	Mycoplasma pneumoniae	Lungs (primary atypical pneumonia)
Tuberculosis	*Mycobacterium tuberculosis*	Lungs, Bones, Other organs.
Botulism	*Clostridium botulinum*	Neuromuscular junction
Food poisoning	*Staphylococcus aureus*	Intestine
	Clostridium perfringens	Intestine
Q-fever	*Coxiella burnetii* (Rickettsia)	Lungs
Typhoid fever	*Salmonella typhi*	Intestine, Blood, Gall bladder.
Cholera	*Vibrio cholerae*	Intestine
Shigellosis	*Shigella* (serotype)	Intestine
Anthrax	*Bacillus anthracis*	Blood, Lungs, Skin
Tetanus	*Clostridium tetani*	Nervous system, Neuro-muscular junction
Gas gangrene	*Clostridium perfringens*	Muscles, Nerves, Blood cells.
Plague	*Yersinia pestis*	Lymph, Nodes, Blood, Lungs
Syphilis	*Treponema pallidum*	Skin, Cardiovascular organs
Gonorrhea	*Neisseria gonorrheae*	Urethra, Cervix, Fallopian tubes
Vaginitis	*Gardnerella vaginalis*	Vagina
Leprosy	*Mycobacterium lepreae*	Skin, Bones, Peripheral nerves
Trachoma	*Chlamydia trachomatis*	Eye
VIRUS		
Influenza	RNA Virus	Respiratory tract
Herpes simplex	DNA Virus	Skin, Pharynx, Genital organs
Small pox	DNA Virus	Skin, Blood

MICROBIAL DISEASES OF MAN (contd.)

Chicken pox	DNA Virus	Skin, Nervous system
Measels	RNA Virus	Respiratory tract, skin
Mumps	RNA Virus	Salivary glands, Blood
Enchephalitis	RNA Virus	Brain
Poliomyelitis	RNA Virus	Intestine, Brain, Spinal chord
AIDS	Retrovirus (RNA)	T-lymphocytes
Cytomegalovirus	DNA Virus	Blood, Lungs disease
Hepatitis A	RNA Virus	Liver
Yellow fever	RNA Virus	Liver, Blood
FUNGI		
Candidiasis	*Candida albicans*	Vagina, Skin
Tinea pedis	*Trichophyton* Sp.	Skin
Tinea corporis	*Epidermophyton* spp.	Skin
Histoplasmosis	*Blastomyces dermatitides*	Lungs, Other organs
Aspergillosis	*Aspergillus fumigatus*	Lungs, Ear.
PROTOZOA		
Amoebiasis	*Entamoeba histolytica*	Intestine, Liver
Giardiasis	*Giardia lamblia*	Intestine
Trichomoniasis	*Trichomonas vaginalis*	Urogenital organs
Leishmaniasis	*Leishmania denovani*	WBC, Skin, Intestine
(Kala-azar)		
Toxoplasmosis	*Toxoplasma gondii*	Blood, Eye
Malaria	*Plasmodium* sp.	Liver, RBC
Sleeping sickness	*Trypanosoma brucei*	Blood, Brain

Depending on the sources of infection bacterial diseases can be broadly categorised as Food and Water borne diseases, Air borne diseases, Soil borne diseases and Sexually transmitted diseases.

Food and Water borne diseases

Some common dreaded diseases of this category are enumerated below :

Typhoid fever

Typhoid fever is caused by *Salmonella typhi,* a Gram-negative rod bacterium. The bacterium is transferred to a susceptible individual by water or food and from infected individuals. Typhoid fever is an infection of the lymphatic system and other tissues. The disease begins with the invasion of mucosal epithelium and rapid movement of the pathogen to lymphoid tissue associated with gastrointestinal tract. The invading pathogens multiply in the lymphoid tissue, move to the blood and spread through out the body. The typical typhoid symptoms include headache, fever, malaise, spleen enlargement and constipation. This ultimately results in the necrosis of Peyer's patches in the intestine producing hemorrhages. If the intestinal wall becomes perforated, the infecting pathogens enter the peritoneal cavity causing **peritonitis** leading to death.

Although infection does stimulate the formation of antibodies, the degree of active immunity produced is not sufficient to prevent a second infection. Some antibiotics like chloramphenicol are to some extent effective against the pathogen.

Botulism

This is the most dangerous of the food-borne diseases. The bacterium responsible for the disease, *Clostridium botulinum* exists in the form of spores in the intestine of many fishes, birds and barnyard animals. Under anaerobic condition these spores germinate and produce exotoxin inside the body. This exotoxin is very powerful that is absorbed readily in the blood stream. Within hours the patient begins to develop paralysing symptoms. Blurred vision, impaired speech, difficulty in swallowing and respiratory distress are the common complications. Death by respiratory paralysis may occur within a day or two.

Antibiotics are not much effective against botulism. Large dose of botulism antitoxin are usually administered to neutralise the effect of exotoxin.

Clostridial Food Poisoning

Clostridium perfringen produces an enterotoxin responsible for food poisoning. Although not as strong as the toxin produced by *Clostridium botulinum*, this enterotoxin acts on the lining of the intestine to cause diarrhoea and nausea that last for one day or less. The poisoning occurs after ingesting foods contaminated by endospores from the soil or fecal material. Death, however, is rare and the disease can be treated by antitoxin and penicillin.

Cholera

Cholera is another diarrheal disease caused by enteric Gram-negative bacterium *Vibrio cholerae*. This bacterium was first of all identified in 1883 by Robert Koch. Transmission of the bacterium occurs through water, contaminated food and from chronic carriers. After the microbes are ingested they undergo an incubation period of two to three days before symptoms are expressed. The bacterium adheres to receptor sites on intestinal epithelial cells and releases a powerful enterotoxin and mucinase. The enterotoxin is responsible for the excessive water loss from the surrounding tissues and inhibition of sodium uptake. Mucinase causes the sloughing off of a surface epithelium. These changes result in severe water loss and an electrolyte imbalance that ultimately leads to diarrhea, dehydration, acidosis, shock and death.

Treatment of cholera victims requires both antibiotic therapy and replacement of lost body fluids. Immunisation with killed vibrio has been attempted with some success but routine sanitory control measure is the best way of preventing the disease.

AIR BORNE DISEASES

The air borne bacterial diseases are mainly of respiratory tract. In crowded population and under poor sanitation these diseases are more frequent.

Tuberculosis

Tuberculosis has been one of the most common communicable diseases, largely prevalent among the people of lower strata of society. *Mycobacterium tuberculosis* var. *hominis* is the causal pathogen responsible for infection in man. It gets entry through the upper respiratory tract. Tubercle bacilli donot release toxins rather they cause disease by their growth and interaction with the host. Although the most familiar type of tuberculosis occurs in the lungs, infection may also occur in gastrointestinal tract and conjunctival membrane and skin. The two recognised forms of tuberculosis are exudative and productive. Exudative infections may occur anywhere in the lung. Upon entering the lungs the tubercle bacilli stimulate migration of PMN, monocytes and macrophages to the focus of infection. If the body's defenses are not sufficient to handle the infection, the bacilli cause massive necrosis of the infection site. The patient discharges bacilli in the sputum. In severe cases the pathogen becomes established in kidney, liver and meninges and the disease is referred to as acute meliary tuberculosis. The productive type lesion may form at the initial pulmonary infection site and is recongnised as a nodule composed of three distinct zones or layer called tubercle. As the tuber develops, the infected tissue takes on a 'cheesy' appearance and results in caseation necrosis. The most common clinical symptoms of pulmonary tuberculosis are fatigue, weight loss, fever, chronic cough and blood in sputum.

Attempts at inducing immunity through vaccination with attenuated *M. tuberculosis* have been made with appreciable success. Attenuated live BCG (Bacille Calmette Guerin) provides protection against tuberculosis especially in childhood. Chemotherapy involves the use of antibiotics that penetrate the tubercle. Rifampicin, ethambutol, streptomycin etc. are the common antibiotics used for the disease.

Diphtheria

Diphtheria or diphtheria pharyngitis is caused by Gram-positive bacterium *Corynebacterium diphtheriae*. The bacterium constitutes a part of the normal flora of skin and mucous membrane and is found in both healthy and diseased individuals. Bacteria enter the upper respiratory tract through droplet nuclei and become lodged on the tonsils. They release a powerful exotoxin capable of inhibiting protein synthesis and causing nerve damage. As the bacteria invade the mucosal epithelium, they destroy host cells and stimulate the formation of thick, firbinous exudate filled with dead cells, leukocytes and fibrins. This tough, gray tissue forms a characteristic pseudomembrane which spreads and may fatally involve the larynx. The difficulties in breathing and swallowing may ultimately lead to suffocation. The exotoxins cause necrosis of the heart, liver, kidney and adrenal gland.

Vaccination of children in combination with tetanus and pertussis toxoid (DPT) is quite effective in preventing the disease. Administration of diphtheria antitoxin effectively neutralise the circulating exotoxins.

Pertussis (Whooping cough)

Pertussis is a very common upper respiratory tract disease of young children. The causal oranism *Bordetella pertussis* is a Gram-negative rod that is transformed through droplet nuclei or diseased individuals. Once inside, the disease passes through three distinct stages : (i) catarrhal (inflammed mucous membrane) (ii) paraxysmal (spasm) and (iii) convalescent. During the catarrhal stage which lasts from one to two weeks, the patient develops symptoms typical of most upper respiratory infections. The release of endotoxins irritates the epithelium to produce a mild cough, sneezing edema and lymphocytosis. In paraxysmal stage which lasts from four to six weeks the patient experiences repeated and uncontrolled coughing. Series of explosiv cough occur in a single exhalation followed by typical whooping sound. The frequency of cough and difficulty in breathing interfere with swallowing. Coughing becomes so frequent and spasmotic that exhaustion, vomitting and convulsions may occur. The third or convalescent stage develops as the frequency of whooping and coughing decreases. Although the fatality is rare, secondary invaders cause complications such as meningitis , pneumonia or influenza.

Pertussis bacilli are susceptible to many antibiotics, including erythromycin, ampicillin and chloramphenicol. The pertusssis vaccine has been quite effective in reducing the incidence of whooping case.

Meningococcal meningitis

This is caused by *Neisseria meningitides,* a small Gram-negative diplococcus commonly called the meningococcus. The disease is spread by droplet nuclei and passes through the mucous membrane of upper respiratory tract into the blood stream. Due to release of large amount of endotoxins, an endotoxin-shock meningococcemia develops. After the blood infection, the bacteria localise on the covering of spinal cord and brain. The condition is commonly known as meninges. It causes severe headache and delirium and death may occur in most of the untreated patients.

Sulfonamide drugs, rifampicin and ampicillin are often used against the disease.

SOIL BORNE DISEASES

Many pathogenic bacteria are the common inhabitants of soil, dungs or debris and they enter into the body through the skin abrasion or wounds causing serious diseases. The common soil borne diseases are :

Tetanus

Tetanus or 'lock-jaw' gets its name from the spasmic contractions (tetani) of muscles in the jaw, neck and face. The symptom is produced by the powerful neurotoxin, tetanospasmin released by an obligate Gram-positive anaerobe *Clostridium tetani*. The spasm usually begins at the site of infection and progresses throughout the body. The incubation period of the disease ranges from two to fifty days and symptoms first appear as headache, stiff neck and minor spasms. The symptoms develop very rapidly and soon attack the jaw muscle causing the dreaded lock-jaw and arching of the back. Death is caused by asphyxia (suffocation) due to contraction of respiratory muscles and blocking of nerve pulse transmission.

It is imperative that the toxin be neutralised as soon as possible by antitoxin antibody. Immunisation with toxoid is by far the most successful therapy.

Gas gangrene

Like tetanus, gas gangrene has also been associated with any injured tissue infected with *Clostridium perfringenes*. The microbe produces subterminal endsopores that may be found in soil, intestinal tract or on the skin and can grow in condition of low oxygen. Once established, the vegetative form releases gases and a variety of toxins and enzymes. Amputation of the infected part becomes desirable to check the spread of the disease. The toxins released by the pathogen spread throughout the body causing heart and kidney destruction and ultimately death.

Besides antibiotics therapy, superficial infection is largely controlled by elevating the oxygen level of the diseased tissue causing the anaerobes either to die or revert to their harmless endospore forms.

SEXUALLY TRANSMITTED AND CONTACT DISEASE

Gonorrhea, syphilis (genital contact) and leprosy (skin contact) are some of the important bacterial diseases that are transmitted sexually or through other contacts.

Gonorrhea

Neisseria gonorrhoeae, the Gram-negative diplococcus is responsible for the venereal disease gonorrhea that occurs only in humans. The pathogen enters the body during sexual contact through the non-ciliated columnar urethral or buccal epithelia. The symptoms of the disease in males are considerably different than in females. Urethral infection in male is characterised by a burning sensation (dysuria) while urinating and the discharge of yellowish pus. The symptoms in females are much less obvious and may be limited to an increased vaginal discharge of yellowish watry fluid. The absence of any major perceptible symptoms enables the female to serve as a main reservoir of infection. Infection may progress through the urogenital system to include the prostrate epididymus and testes in males and the fallopian tubes and ovaries in females leading to sterility. In some cases the bacterium enters the blood and become localised in the joints causing arthritis or in the heart resulting endocarditis or in the skin where small red pustules are formed. In some females the disease becomes chronic and results in an asymptomatic cervicites.

Since the pathogen is intracellular parasite, the immue symtem has little effect. Streptomycin with tetracylin is used to restrict the disease.

Syphilis

The disease is caused by *Treponema pallidum,* a spirochete that moves by axial filaments. Sexual contact is the most common form of disease spread. The first symptom of the disease is the appearance of chancre at the point of entry of spirochete, mainly on the external genital organs. The lesions may appear on the skin, lips, rectum and pharynx. Soon wart-like lesions filled with spirochetes appear over the entire body surface. The disease remains dormant for many years and finally soft granular lesions called **gummas** develop resulting in paralysis.

Penicillin may control disease satisfactorily at primary and secondary stages.

Leprosy

The disease is caused by a heat sensitive organism *Mycobacterium leprae*, which invades the nerves and cutaneous tissues causing degeneration and deformation of the body parts. There are two forms of leprosy :

(i) Lepromatous leprosy - it is characterised by tumor-like growth called lepromas on the skin and along the respiratory tract; skin may lose pigmentation

(ii) Tuberculoid leprosy - it involves the superficial nerve allowing muscular atrophy, disfiguring of skin and bones, twisting of limbs and curling of fingers to form characteristic claw hand. Death occurs mostly due to secondary infection.

Treatment of disease in initial stages effectively cures the ailment. Antibiotics as rifampicin, clofazimine and sulfone drugs are quite effective against the disease.

VIRAL DISEASES

Viral diseases are classified according to the nature of their nucleic acid core into two major groups: the DNA viral diseases and the RNA viral diseases. Further subgrouping is based on the morphological, immunological and epidemiological differences. Some of the common viral diseases are as follows :

DNA VIRUS

PARVOVIRUS

This helical, single stranded virus has not been recognised as human pathogen. It primarily attacks the domestic animals like dogs and is accordingly known as canine parvovirus. The symptoms of the infection are more acute in the puppies that include loss of appetite followed by severe diarrhea. The animal gets rapidly dehydrated in 24 to 48 hours resulting in death.

PAPOVA VIRUS

The name papova virus is derived from the first two letters of three oncogenic (see p. 148) Prototypes: rabbit *papilloma* virus, *polyoma* virus of mice and *siman vacuolating* virus (SV40) of hamster and mice. One member of this group of human importance is the human wart virus. This virus causes a benign outgrowth of the skin typically occurring on the feet, hands and other parts of the body. Although painful and time consuming to remove, no significant harm results from the growth of the infected malpighian layer of the skin. Sometimes malignancy develops causing tumor of skin.

POX VIRUS

Viruses of this group are responsible for small pox, vaccinia, cowpox, monkey pox etc. The name 'pox' is derived from the typical eruptive lesion formed during the course of infection. Many pox viruses produce recognisably different symptoms but are immunologically related. They are amongst the largest of the animal viruses. The most significant human pox virus is small pox (variola) which has been a scourge throughout the world. The virus is acquired by close contact with infected individual or contaminated articles. Susceptibility is universal and mortality rate is high. The two forms of diseases are variola major (malignant small pox) and variola minor. These two forms follow the same course and differ only in severity. The virus enters through mucous membrane of the upper respiratory tract and moves possibly through the lymph system and becomes a primary viremia, once it enters the blood. Small pox virus enters the viscera (e.g., spleen, liver) where it multiplies during its 7-17 days incubation period. The release of replicated virions into the blood produces secondary viremia producing characteristic symptoms of chill, fever, vomitting and development of pox on the skin, mucous membrane and viscera. In case of variola major the pox changes from a mascular (thickened, stained spot) lesions to a pustular lesion causing haemorrhaging and necrosis into the surrounding tissue.

Immunity resulting from infection is strong and long lasting. As a result of world-wide immunisation campaign the disease has been considerably eliminated.

HERPES VIRUS

This group of DNA viruses are able to infect many animals including humans. The four forms responsible for human diseases are Herpes simplex virus (e.g., cold sores), Cytomegalo virus (e.g., salivary gland disease), Varicella-Zoster (e.g., chicken pox and shingles) and Epstein-Barr virus (e.g., infective mononucleosis).

Herpes Simplex Virus

Commonly known as HSV, the group is comprised of many polyhedral viruses that are broadly subdivided into two major groups : HSV-I and HSV-II. These viruses are transmitted by oral-respiratory and uro-genital routes and symptoms follow a general pattern of primary and secondary infection. Primary infection is first noted as burning sensation to a localised cluster of nodules. They progressively fuse to form a large, thin walled vesicle filled with fluid, viscous and epithelial cells. The vesicle ruptures and heals without scarring. The infection results in cold sores of the lip and may involve adjacent mucous membrane. HSV-II infection typically occurs at the time of puberity; sometimes in infants born to infected mothers. The disease is highly contagious when sores are present and may be transmitted through sexual contact.

Cytomegalo virus

Cytomegalo inclusion disease (CID) occurs most frequently in infants and is a viral disease of salivary gland. It is caused by the intra-uterine (congenital) or post natal transmission of cytomegalo virus (CMV). Like other herpes viruses, CMV infection results in vesicular eruption in the host tissues. The virus may also infect the kidney, liver, brain, lungs and eyes. Tissue destruction can be fatal or results in severe brain damage, blindness, deafness and heart defects. At present there is no treatment of this disease.

Varicella - Zoster Virus

This is a single virus responsible for two human diseases, varicella (chicken pox) and zoster (shingles).

Chicken pox is highly contagious skin disease occurring in children and adults both. The most likely transmission route is by direct contact, droplet nuclei or contact with contaminated articles. Once inside the body there is an incubation period of two to three weeks before the development of a maculopapular rash which changes to a vesicular form in three or four days. The vesicles become pustular, eruptive, leaving no scar of the scab. Children with immuno-deficiency are likely to experience more severe symptoms. Secondary infection like Reye's syndrome involving degeneration of liver, kidney or other organs may develop in some cases.

Zoster (shingles) typically occurs in adult individuals who have experienced varicella in childhood. In most cases the disease appears on torso and neck resulting in vesicle formation. Since the majority of varicella-zoster infections are mild and self-limiting there is no need of treatment. Chicken pox vaccine is not yet available.

Epstein - Barr Virus

Early EBV infection of children results in mild, almost unrecognisable case of infections mononucleosis. Symptoms include nausea, vomitting, fever, loss of appetite, sore throat, fatigue and most typically tender enlarged cervical lymph nodes. The liver and spleen become enlarged and susceptible to rupture. Infection mononucleosis is considered by many to be a self-limiting form of leukaemia since during the course of the disease there is marked increase in the white-blood cell count (i.e., B-cells). The disease is also known as the 'kissing-disease' since it is thought to be transmitted by the oropharyngeal route. EBV has also been strongly implicated as the cause of Burkitt's lymphoma and nasopharyngeal cancer.

ADENO VIRUS

The adenoviruses are commonly found as latent infections of the adenoid and tonsillar tissue. In human three common diseases are known to be caused by this virus - acute respiratory disease, epidemic kerato-conjunctivitis and pharyngo-conjunctivital fever. Acute respiratory disease of the mucous membranes displays fever, phryngitis and cough. The pharyngo conjuctival fever is characterised by fever, conjunctivitis, fatigue and swollen cervical nodes. No corneal damage results from this short-lived illness. In keratoconjunctivitis there is an enlargement of the auricular lobes (behind the ears), lymph nodes and clouding of the cornea resulting in impairment of vision.

RNA VIRUS

Common RNA viruses which initiate diseases are picorna viruses, rhabdo viruses, hepatitis viruses, paramyxo viruses, arbo viruses and orthomyxo viruses.

Picorna Virus

This group is divided into the enteroviruses and the rhinoviruses. One of the best known members of the enterovirus group is the *Polio virus*.

Polio virus

Poliomyelitis is prevalent throughout the world. Though popularly known as 'infantile paralysis', adults are also equally affected by this disease and sometimes more severely than the children. The disease is acquired through the oral-respiratory route and causes fever, constipation, headache, nausea, vomitting and stiffness of the back and neck. If the virus enters the circulatory system, an asymptomatic viremia occurs that can lead to paralytic poliomyelitis. Symptoms begin after the virus has infected and lysed motor neurons of central nervous system. If the infection progresses to the spinal cord and brain stem, paralysis may be fatal.

Currently no treatment of the active disease is known. Prevention of the disease has been achieved through the use of the Salk and Sabin vaccine.

Cox-Sackie Virus

Recently a cox-sachie virus called B4 has been implicated as a cause of juvenile - onset diabetes. Children with the disease can suddenly pass through a comatose stage in only a few days of infection and die if untreated. Factors influencing susceptibility to this infection have not yet been determined. Researches are being conducted for the development of a vaccine against this virus.

Rhino Virus

This virus has been isolated more frequently than any other viruses from the 'nose' (*rhino* = nose) and throats of the people with the common cold. Over 100 of these RNA viruses have been serotyped and are known to cause edema, excessive nasal secretions and localised tissue sloughing-off of the submucosa. The viruses are transmitted through the droplet nuclei and the incubation period lies between two to four days. Headache, bodyache, watery eyes, chilling, sneezing, malaise are the common experiences of the infection.

The possibility of developing a vaccine for common cold is currently being exploited. However, difficulties in culturing rhinovirus and the role of antibody IgA providing short-term immunity have hindered the venture on experimental state. Use of interferon, **propanediamine** has been found effective experimentally for prevention of the disease, if administered before the exposure of the virus.

Rhabdo Virus

Rabies is known worldwide as a viral disease of major human importance. It is a bullet-shaped RNA virus with spikes and envelope that is transmitted by wild animals and rabied dogs. The virions enter a susceptible host through breaks or open wounds in the skin caused by animal bites. Clinically victims of the rabies display fever, headache, tingling sensation around the wound followed by alternating period of depression and excitability. As the disease progresses and the period of frenzy increases the patient experiences painful spasmodic contractions of the throat muscles; persons suffer from hydrophobia (fear of water). Difficulty in swallowing leads to drooling (foaming at the mouth), frenzied convulsion, comatose and death .

The effective use of vaccine to control and prevent the disease is made possible. Fourteen inoculation of DEV (duck embryo vaccine) are given abdominally to ensure neutralisation of the rabied effect and modified HEP (high egg passage vaccine) is used to immunise the domesticated animals.

Hepatitis Virus

The hepatitis virus group is composed of several complex viruses able to cause acute infection of liver. There are three different subgroups known as hepatitis type A virus (HAV), hepatitis type B (HBV) and non-A non-B hepatitis virus (NANB). HAV is commonly known as 'infection hepatitis' which is a typical water-borne infection that manifests the characteristic symptoms jaundice. Most patients recover with difficulty and develop a strong active immunity. HBV is known as 'serum hepatitis' and is most frequently transmitted through blood, blood products or on blood contaminated instrument. Recent evidences point out that the patients who have experienced HBV infection also run a higher risk of liver cancer. NANB is a recently discovered disease among individuals who hae received transfusion of blood. This liver infection also closely resembles HBV clinically but little is known about this new form of hepatitis.

Paramyxo Virus

This group contains some of the largest enveloped RNA viruses that causes many common diseases like para-influenza, mumps and measels. The para influenza virus demonstrates symptoms similar to that of common cold, finally leading to pneumonia and bronchitis. The mumps virus is highly contagious especially among the children that infects the paratid salivary gland. The virus is spread by droplet nuclei or direct contact with saliva from infected individuals. The incubation period ranges from 12 to 26 days displaying the apparent symptoms of fever, swelling and tenderness of salivary glands. Some rare complications of mumps infection are aseptic meningitis and pancreatisis. Like mumps, measels is also spread by droplet nuclei or direct contact with nasal and other secretions from the infected individuals. The incubation period in this case lasts about ten days before the onset of fever, muscle pain, photophobia (abnormal sensitivity to ligh) followed by measels rash on the skin, buccal mucosa and viremia. Rubella or German measels is very common in children and adults alike. This virus can cross the placenta and cause severe congenital malformations (cataracts, blindness, septal defects, brain retardation etc.) if infection occurs during the early stage of pregnancy.

Immunisation against measels is practised now a days alongwith the combined vaccine of attenuated mumps and rubella virus.

Arbo Virus

These viruses are transmitted and multiplied in arthropods which serve as both hosts and vectors. The three major subgroups of arboviruses are togaviruses, bunyaviruses and arboviruses. Members of these groups are responsible for three types of human illness : (i) encephalitis, inflammation of the brain leading to degenerative changes that may be fatal (ii) hemorrhagic fever, involving extensive external and internal disease with acute fever and high fatality (iii) febrile illness, demonstrating high fever that lasts for a short duration.

Togaviruses are mosquito-borne (*Culex* and *Aedes*) and all cause encephalitis. The most familiar forms are Eastern Equine Encephalitis (EEE), Western Equine Encephalitis (WEE), Venezuelan Equine Encephalitis (VEE), St. Louis Encephalitis (SLE) and Japanese Equine Encephalitis (JEE). Once the infection is localised in the brain, spinal cord and meninges severe degenerative changes occur resulting in mild, almost asymptomatic condition that could be fatal within twenty four hours. Symptoms include high fever, tremor, convulsions and coma.

Yellow fever is a hemorrhagic disease transmitted by infected mosquito *Aedes aegypti*. Like encephalitis the symptoms may be mild or severe and occur suddenly. Fever, headache, nausea and vomitting occur in mild case whereas in acute phase which occurs three or four days later, kidney, liver, spleen and heart get infected and weakened leading to death. Calorado tick fever is a febrile illness which is acquired from the bite of the tick *Dermacentor andersoni*. The patient experiences fever, headache, bodyache, loss of appetite and occasionally a rash that may disappear in three or four days.

No effective immunisation against these diseases are known so far.

Orthomyxo Virus

Influenza viruses are the members of this important group. They are small, enveloped virions each containing several strands of RNA. Viruses have a ribonucleoprotein core, the outermost cover containing two important antigens, hemaglutinin antigen (HA) spike and neuraminidase antigen (NA) spikes. Based on the antigenic spikes, the influenza viruses are classified into three major antigenic types. Type A influenza viruses are more frequently involved in flu and undergo frequent antigenic changes known as antigenic drift. Type B is antigenically more stable than type A and has rarely been implicated in epidemics. Influenza type C is considered to be a stable antigenic form that occurs sporadically. Influenza viruses are easily transmitted by droplet nuclei and remain localised on the lining of upper respiratory tract. No viremia develops. Symptoms of infection include chill, muscle pain, malaise, heavy mucous discharge, fever and prostration that last from one to two days.

Immunity to the disease develops shortly after the disease but is short-lived. The short life of IgA and the rapid antigen drifts have made it difficult to develop flu vaccine.

Besides the above common viral diseases of human beings, two dreaded diseases of recent origin - AIDS and Cancer - have received considerable attention throughout the world. A brief mention of these diseases are given below :

Acquired Immune Deficiency Syndrome (AIDS)

One disease, one fight, hundred of governments, thousands of dedicated professionals ... that is the story of AIDS today. Since the first report of AIDS in 1981 from USA millions of people around the world are suspected to be the victims of the disease that has already killed more than 35,000 people till date. The disease is quite prevalent in America, Australia, South and Central Africa , different parts of Europe, the Carribean and Japan. Reports of the incidence of AIDS are also appearing from various parts of India. Suvey data reveals that more than 80 per cent of the suspected patients of the world are in USA alone that constitutes 92.5% male, 6.5% female and 1% children. Majority of the patients are homosexual (75%) and intra-venous drug users (15%).

Though AIDS is largely confined to male homosexuals, good percentage of heterosexuals are also the victims of this disease. The spread of disease is mainly through the blood and semen fluids of the infected persons. The risk of contracting the disease increases considerably living in high risk areas sharing with (i) homosexual (ii) intravenous drug abusers (iii) female partners of any i & ii (iv) transfusion of whole blood plasma or platelet and (v) unsterilised syringes and needles. Transmission of disease through infected mothers has also been reported. Patients of high risk areas are the healthy carriers of the virus (first stage of disease exhibiting no apparent symptom. They remain so throughout their lives and are capable of infecting others. On prodromal stage (2nd stage) the patient experiences fever, diarrhoea, night-sweat, weight loss and enlarged lymph nodes. The end stage is a full blown picture of AIDS. A number of uncontrollable infections appear, the immune system completely collapses and the life span of the patient is drastically cut short to one to four years. Data reveals that majority of AIDS patients develops one or both of two rare diseases Pneumocystis carini pneumonia (PCP), a parasitic infection of lungs; and a type of skin cancer known as Kaposi's sarcoma (KS) caused by cytomegalo virus. The overall symptoms of AIDS disease are : dry cough breathlessness, and vague chest pain and discomfort. Neurologically the patient develops loss of memory and convulsions and in advanced stage malignancy of lymph nodes (Non-Hodgkins B-cell lymphoma) and carcinoma of mouth and rectum develop leading to comatose and death.

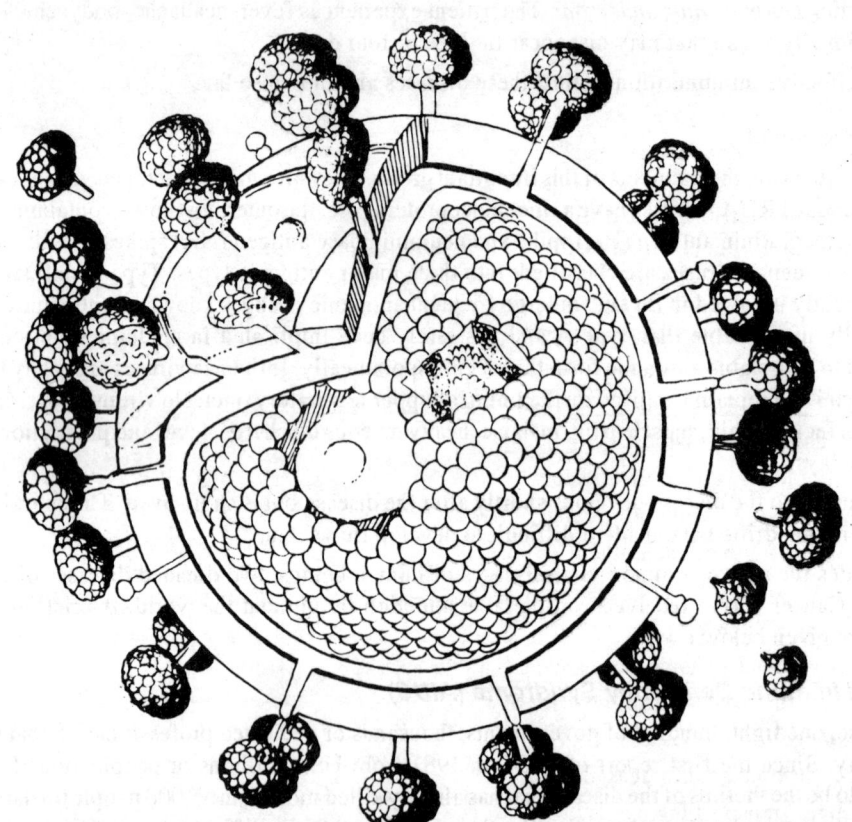

Fig. 10.3 The AIDS virus.

AIDS is caused by a Retrovirus that has been given different names by different investigators, such as human T-lymphotropic virus type III (HTLV - III), Lymphoadenopathy associated virus (LAV) or AIDS-related Retrovirus (ARS). An International Committee of virologists has proposed a name Human Immuno deficiency virus (HIV) for this pathogen. The virus has a liking for T-lymphocyte-T helper (T_4) lymphocyte causing a decline in the number of helper T-cells and a change in T_4:T_8 (suppressor) lymphocyte ratio. The iberated virus particles enter the fresh lymphocyte to repeat the cycle till all lymphocytes are killed. Thus he whole immune system collapses and the patient passes into a defenseless state. AIDS kills ruthlessly, deceptively, remorselessly and surely. (Fig. 10.3)

Drug therapy and vaccine against the disease are so far, not successful. A number of candidate drugs ike AZT (azidothymidine), Retrovir, Peptide T have been released in the market. α-interferon γ-interferon, interleukin-II, Thymopentin, Thymostimulin etc. are also on the trial stage in an attempt to tame the AIDS virus.

Cancer

Malignant tumors are known as cancer. A cancerous cell escapes the normal regulatory processes and it divides without restraint forming an abnormal mass of tissues. Such masses are called tumors or neoplasms. Some of the tumors are localised and harmless (benign tumor) while others are invasive, affecting the organs (malignant tumor). Tumors are usually named by adding the suffix 'oma' (= forming tumor) to the name of the tissue. For example, lymphoma is the cancer of lymphoid tissue, adenocarcinoma is the cancer of glandular tissue, sarcoma is the cancer of connective tissue (the only exception is leukaemia - cancer of white blood cells, the connective tissue). The cancer causing agents are known as carcinogens which may be physical (UV, ionised irradiation), chemical (mustard gas, benzidine, aflatoxin) or biological (oncogenic virus) in origin.

Although cancer is a disease definable only in whole animals, an analogue malignancy called **cellular transformation** provides the *in vitro* model for the detailed study of cancer. The transformed cells exhibit certain abnormal characteristics quite distinct from the normal cells (Table-10.3). Some common manifestations of cancerous state as evidenced by transformed cell culture are as follows :

Table 10.3
CHANGED PROPERTIES IN TRANSFORMED CELLS

Morphological and Behavioural Changes -

* Become more rounded.
* Have looser attachment to substratum.
* Mutual orientation more random; lose contact inhibition of movement.
* Grow on top of each other.
* Grow to high or indefinite saturation density.
* Become invasive.
* Grow in suspension, lose anchorage dependence.
* Have decreased serum requirement.

Surface alteration

* Hyaluronic acid increased.
* Sugar transport increased.
* More easily agglutinated by plant lectins.
* Surface protein more mobile.
* Lipid fluidity not changed.
* Microfilaments disappear, microtubules disaggregate.
* Fetal antigens become evident.

Non-surface biochemical changes

* Release of protease.
* Transcription of fetal genes.

Immortalisation

Primary or secondary cell cultures have a finite life expectancy. Human cell culture, for instance, die about 50 cell generation; chick cell cultures have a shorter life span. By contrast, most transformed cells are immortal; they will grow indefinitely.

Increased saturation density

Most normal cells grow adherent to substratum. They stop growing before they have exhausted the nutrients and then remain in quiscent viable state. These cells form a monolayer on the surface of culture dish because their nuclei do not overlap. In sharp constrast, most transformed cells will grow continuously until they kill themselves. The continuously growing cells also do not respect each other borders and grow chaotically over and under each other forming multilayered masses cells causing malignancy.

Sensing mechanism

Normal cells can sense each others when they come into contact. As soon as they approach each other one or both the cells will stop moving and then separate in opposite direction. This phenomenon is known as **contact inhibition movement**. Transformed cells lack such sensing mechanism and so grow on top of each other.

Anchorage dependence

Normal cell is attached to a rigid substratum whereas transformed cells can grow without such attachment. Such loss of anchorage dependence is correlated with the ability for **metastasis**.

All such changes lead to a complex of several diseases and the victim is destined to slow but certain death.

So far the involvement of viruses to the disease is concerned there is convincing evidence that viruses are associated with a number of specific cancers in animals (Table-10.4). However, there is no clear cut evidence of malignant tumors being caused in human by virus. Yet, the associations of EB virus with Burkitt's

Table 10.4
SOME IMPORTANT ONCOGENIC VIRUSES

	Virus	*Cancer in Host*
DNA Virus	Papilloma virus	rabbit
PAPOVA VIRUSES	Polyoma virus	mice
	Simian vacuolating virus (SV40)	tumor in hamster and mice.
ADENO VIRUSES	Adeno 12, 18 and 21	malignancy in hamster and rats.
HERPES VIRUSES	Epstein-Barr (EB) virus	infection mono-nucleosis and Burkitt's lymphoma in children.
	Herpes simplex Type-2	virus Carcinoma of uterus, cervix
	Lucke's disease virus	Adenocarcinoma (renal tumor in frogs.
	Marek's disease virus	Tumor in birds. virus
POX VIRUSES	Molluscum contagiosum virus	Benign epidermal tumor in man.
	Yaba monkey tumor virus	Subcutaneous tumor in monkey and man.
RNA VIRUSES		
RETROVIRUSES	Rous sarcoma virus (RSV)	Sarcoma in chicken.
	Murine Sarcoma virus (MSV)	Sarcoma in mice.
	Murine leukaemia virus (MLV)	Leukaemia in mice.
	Avian sarcoma virus (ASV)	Sarcoma in birds.
	Avian leukosis virus (ALV)	Leukaemia in birds.

lymphoma or herpes simplex type 2 virus with cervical cancer or virus-like RNA-DNA particles with breast cancer and leukaemia present ample evidence that viruses do play a role in human cancer.

Of the DNA tumor viruses (papovaviruses, adenoviruses, herpes viruses and pox viruses), the papova viruses have received considerable attention because they are known to cause tumors in a variety of animals. Retroviruses are the only RNA viruses (popularly known as **oncornavirus**) responsible for leukaemia and carcinoma. It was first discovered by F. Peyton Rous in 1911 from the connective tissue of chickens causing Rous sarcoma, for which he was honoured with Nobel Prize in 1960. Oncogenic RNA viruses have been grouped in three classes, A, B and C. Out of these **type cRNA** viruses are the most important oncogenic form causing leukaemia, lymphoma and sarcoma.

How viruses trigger cells to become tumor cell has been the subject of intensive research in the recent past. There are three current viruse - cancer hypotheses based on the assumption that viruses do, infact, cause human cancer.

1. *Provirus Hypothesis*

It was proposed by H.M. Temin in the early 1960's. The theory suggests that after infection by an RNA tumor virus, the cell makes a DNA copy of the viral RNA and incorporates this genetic information into its own DNA. This reaction gives the cell the capacity to produce neoplastic cell from a normal cell. The enzyme *Reverse transcriptase* is supposed to play a key role in the process.

2. *Oncogene Hypothesis*

It was proposed by R.J. Huebner and G.J. Todaro in 1969. It suggests that genetic information for cancer is inherent in every cell in the form of oncogene and is transmitted from parent to child. This oncogene is normally in the repressed state. When a genetic change occurs either by a viral infection or by a chemical carcinogen or by radiation it becomes active and transforms the normal cell to tumor cell.

3. *Protovirus Hypothesis*

Temin proposed this hypothesis in 1970. In many ways it is similar to oncogene theory. The hypothesis holds that cancer virus arises from segments of genetic information randomly brought together by a variety of cellular and genetic events. These segments are termed as *proto-oncogenes*. These may have important functions in the normal regulation of cell growth and development. Mutagenesis by carcinogenes converts the proto-oncogenes into oncogenes that transform the normal cells into malignant cells.

All these hypotheses, however, have certain limitations and the prime concerns of the oncologists have centred more on inventing a reliable remedy of the disease than to unravel the complicacies associated with the origin of the disease.

SUGGESTED READINGS

Baltimore, D. (1979). Viruses, polymerase and cancer. Science, **192**: 632-636.

Briody, B.A. and Grill, R.E. (1974). Microbiology and Infectious disease. Mc Graw Hill, New York.

Cairns, J. (1975). The cancer problems. Scientific American, **232**: 64-77.

Gross, L. (1970). Oncogenic viruses. Pergamon Press, Oxford.

Maugh, T.H. (1974). What is cancer? Science, **183**: 1068-1069.

Ross, F.C. (1983). Introductory Microbiology. Charles E. Merrill Publishing Co., Columbus, Toronto, London.

Schlessinger, D. (ed.) (1975). Microbiology. American Society for Microbiology, Washington D.C.

Stanier, R.Y; Adelberg, E.A. and Ingraham, J.L. (1977). General Microbiology, 4th (ed.) Macmillan, London.

Top, F.H. and Wehrle, P.F. ed. (1976). Communicable and Infectious Disease (8th ed.). The C.V. Moslav Co.. St. Louis.

Tortora, C.J; Funke, B.R. and Case, C.L. (1986). Microbiology : An introduction (2nd ed.) The Beryamin Cummings.

IMMUNOLOGY, HYPERSENSITIVITY AND SEROLOGY

In chapter 10, we have discussed the non-specific resistance in host involving all the protective mechanisms offered by surface layer, anti-microbial secretions, phagocytosis, inflammation etc. In this chapter we shall consider the specific resistance (immunology) acquired in the host in response to a particular parasite (antigen).

Immunity (*immunis* = exempt) has been recognised as a specific host defense mechanism against body's exposure to specific foreign substance and is directed towards a single pathogen or toxin. Until the mid of the 20th century specific resistance to disease was virtually considered to be synonym to immunity. However, with the rapid progress in the field of biological sciences after World War II it became apparent that specific resistance had some wider applications. It has now included many events of cellular metabolism like rejection mechanism, allergic reaction, serological tests, vaccine research etc. within its framework.

BASIC CONCEPT OF IMMUNE SYSTEM

Human immune system responds to a foreign agent in two ways : innate and acquired immunity. The innate or **inborn immune system** allows the host to resist a particular pathogen without having been previously exposed to the infectious agent. No antibody production is involved in this form. Basic physiological, genetic, biochemical and anatomical differences in species are responsible for defending the organism against the pothogen. For example, Chinese are more resistant to venereal diseases as compared to people in the western countries. Diseases of warm blooded animals seldom occur in cold blooded animals and even within the warm blooded animals cross-infection (like birds and mammals) is very rare, probably due to the differences in body temperature.

Acquired immunity is achieved when individual becomes relatively or temporarily resistant to an infectious agent either by natural or artificial means. Depending on the nature of infection, acquired immunity may be active or passive (Table. 11.1). **Naturally acquired active immunity** develops after an attack of the pathogen and subsequent antibody formation in the host. The protection provided by this exposure is usually long lasting and specific (e.g; mumps virus infection). **Naturally acquired passive immunity** is provided to infants by the transmission of maternal antibodies through the placenta or in the colostrum of breast milk. Protection provided by these antibodies is temporary (lasting for 4-6 months). Artificially acquired immunity may also be achieved both in active or passive way. Vaccination with attenuated microbes or with toxoid stimulates the immune system providing **artificially acquired active immunity**. This armors the individual with a long lasting specific defense. *Artificially acquired passive immunity* may be obtained by the injection of antibody previously formed in another organism.

Table - 11.1

THE GENERAL IMMUNE SYSTEM

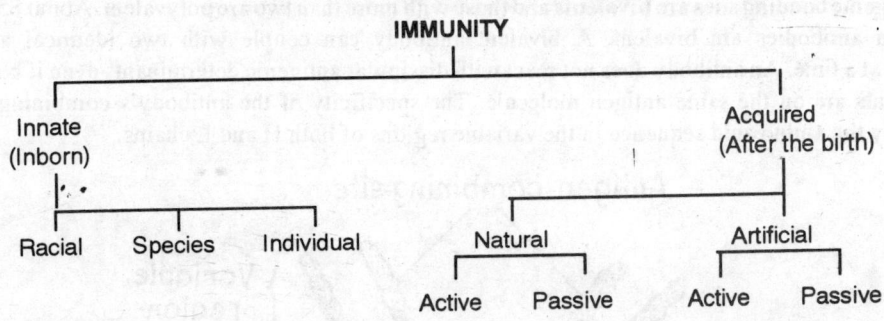

The injection of this antiserum provides immediate protection (e.g; gammaglobulin from horses, cows or rabbit) but is effective only for a short duration because the body has not been stimulated to manufacture its own antibody.

ANTIGEN : Immunity is based on two essential components : antigen and antibody. Any substance that specifically stimulates an immune response when introduced into the body is an **antigen**. Most of the antigens are high molecular weight protein or polysaccharide. Some are soluble (e.g., bacterial exotoxins), others are particulate (e.g., structural components of bacterial cell). H-antigen of flagellum, O-antigen of LPS side chain, M-antigen of peptidoglycan, K-antigen of capsule, cytoplasmic membrane antigen, pilus antigen - all are particulate in nature. Each antigen contains chemically distinct sites called **antigenic determinants** that defines its specificity. These are the molecular regions against which the immune response is directed. In general larger molecules of antigen contain more intense immune system. Sometimes small molecules that are non-antigenic may become antigenic determinants when coupled to a larger carrier molecule. These are called **haptens**.

Antigens must be recognised by the host as *foreign*, before stimulating an immune system. It is quite likely that a substrate that is antigenic in one host may be completely accommodated in another host. Kidney transplant, for example, is rejected in most of the cases because the proteins of the transplanted organ which are recognised as 'self' by the donor, are foreign to the recipient and attacked by the recipient immune system. The mechanism that prevents us from rejecting our own tissue is called **immunological tolerance**. Depending on the type of reactions the antigens can be broadly classified as: *autoantigens, alloantigen* and *heterophile antigens*. Autoantigens are person's own chemical substances which elicit an immune response when self-tolerence breaks down. This may result in an autoimmune disease (e.g; rheumatoid arthritis). The AB and Rh antigens of various blood groups in human are the typical examples of alloantigens. If an Rh-negative woman bears an Rh-positive child, a condition called hemolytic disease of the infant (Rh disease) may develop. Heterophile antigens are identical antigens found in apparently unrelated species of organism.

ANTIBODY : An antigen (Ag) is able to stimulate production of a specific antibody (Ab). An antibody is a protein molecule that represents about 17 percent of the Bence Jones protein (blood serum protein - named after Henery Bence Jones). It is also called as **immunoglobulin** (Ig). Details of structure and specificities of antibody were provided by Gerald M. Edelman and R.M. Porter who received the 1972 Nobel Prize in medicine for their contribution.

Anibodies are monospecific molecules; they combine only with the single type of antigenic determinant that stimulated their formation. Structurally, most antibodies consist of four protien chains linked together (Fig. 11.1) in a 'slingshot' or Y-shaped configuration. The two shorter chains called **light chains** (L) are covalently linked to the branches of the longer **heavy chanis** (H). Each chain has a variable and constant regions. The portion of antibody that chemically binds to a specific antigen is known as the **antigenic binding site** which is usually localised at the top of each arm of the Y. The number of binding sites on an antibody is referred to as the **valency** of the molecule which influences the antigen-antibody interaction. Antibodies with two antigenic bonding sites are **bivalents** and those with more than two are polyvalent. About 85 percent of all human antibodies are bivalent. A bivalent antibody can couple with two identical antigenic determinants at a time. An antibody does not react with dissimilar antigenic determinants even if both types of determinants are on the same antigen molecule. The specificity of the antibody's combining sites is determined by the amino acid sequence in the variable regions of both H and L chains.

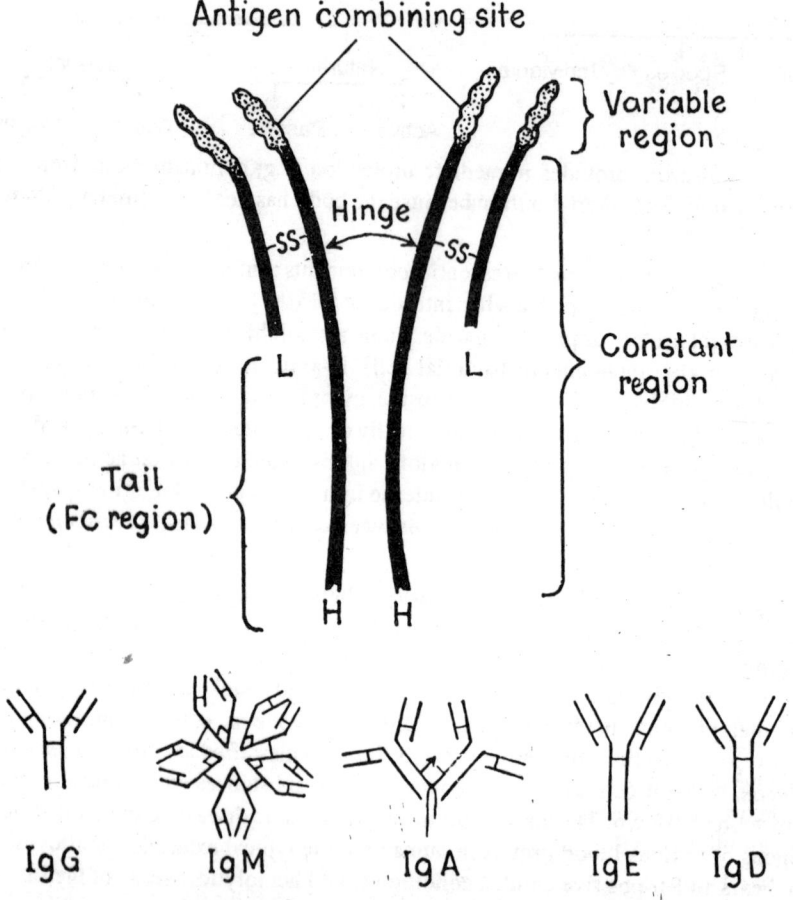

Fig. 11.1 (Top) Structure of an antibody
(Below) Configuration of different forms of antibody.

Based on the properties of constant region (heavy chains), antibodies are grouped in five major classes. These five classes of antibodies have been named after the Greek alphabets and are typically referred to in 'short hand' notations' like IgM (*mu*), IgG (*gamma*), IgA (*alpha*), IgE (*epsilon*) and IgD (*delta*). Properties of five immunoglobulin classes are presented in table 11.2.

Table 11.2

CHARACTERISTICS OF SERUM IMMUNOGLOBULINS

Immunoglobulin class	Molecular weight	Percent in blood	Valency	Location
IgG	150,000	75-80	2	Blood and tissue fluid
IgM	900,000	6-7	5-10	Blood and tissue fluid
IgA	170,000	15-21	1-2	Saliva, mucous or more and other secretions
IgE	200,000	< 1	2	Skin, respiratory tract
IgD	180,000	< 1	2	Serum.

IgG is the most common class of antibody. It is typically Y-shaped, bivalent and can cross placenta, so a mother IgG antibodies help protect her developing fetus. When coupled with antigen it fixes a group of serum protein called **complement** (see p. 154)

IgM antibodies consist of five or more Y-shaped subunits linked together by their tail called *Fc* region. These are the first antibodies to appear in the circulation after stimulation of *B*-lymphocyte and are the principal components of **primary antibody response**. They fix complement but can not cross the placenta.

IgA antibodies are present in two forms. In the serum IgA structure resembles that of IgG and is called **serum IgG**. The second form the **secretary IgA** is the principal antibody found in saliva, mucus and other external secretions. Secretary IgA is a dimer composed of coupled IgA and provides resistance in the respiratory and gastrointestinal tract. It is also located in tears and in the colostrum (breast milk) of the nursing mother.

IgE attaches by its *Fc* region to certain host cells leaving its antigen - combining sites available for binding with antigen. It plays a major role in allergic reaction. Its protective function is unkonwn.

IgD is the least understood type of antibody. It is believed to stimulate B-lymphocyte for the differentiation of immune cells.

ANTIBODY - ANTIGEN INTERACTION

In order to ensure specific resistance against an antigen, the antibody must alter the antigen, in a way so that it is inactivated, killed, prevented from spreading in the body or is rendered more susceptible to other defense mechanism. Five general categories of antibody-antigen reactions have been recognised.

(1) **Agglutination** : An agglutination reaction occurs when an antibody combines with an antigen, that is a part of a large insoluble particle to form an insoluble aggregate of material. Many of these reactions involve bivalent or polyvalent antibodies which may inter-link to form complex arrangements. These complexes are more easily recognised by phagocytes and are quickly engulfed and destroyed. The identification of an ABO blood type is probably one of the familiar agglutination reaction.

(2) **Precipitation** : Unlike agglutination, **precipitation reactions** involve a small, soluble antigen. Precipitation reactions take place over a long period and are less visible than agglutination. As a result of this reaction a few small soluble lattice complex is developed that ultimately results in precipitation. Maintenance of an optimum balance between antibody and antigen is important for efficient operation of the immune system. In human body this maintenance of optimum ratio is largely accomplished by T-cells.

(3) **Lysis** : The third type of antibody-antigen reaction results in the **lysis** of microbial cells or red blood cells. The series of reactions that ultimately create holes in the plasma-membrane begins with the

attachment of free antibody to cell surface antigen and is followed by a complex series of reactions involving 11 serum protein, collectively known as **complement system** (C system). This system was originally described by Joules Bordet in 1895. The system exists in all the normal sera and is activated by IgM or IgG. Coupled with these antibodies the complement system works to cause opsonisation, chemotaxis and lysis of the cell.

Components of the C system are identified by numbered letters, C_1, C_2.....C_9. The additional components that occur for 11 elements are subcomponent of C_1 and are designated as C_{1q}, C_{1r} and C_{1s}. ($C_1 = C_{1q} + C_{1r} + C_{1s}$). Complement fixing antibodies are especially operative in Gram-negative bacteria. These antibodies combine with their antigenic target before the inactive complement components sequentially aggregate into their active forms on the antigen-antibody (Ag-Ab) complex. Complement proteins alter the cell membrane facilitating the leakage of cytoplasm and lysis of cell. The sequential events involving the complement system for the lysis of bacterial cell is presented in figure 11.2.

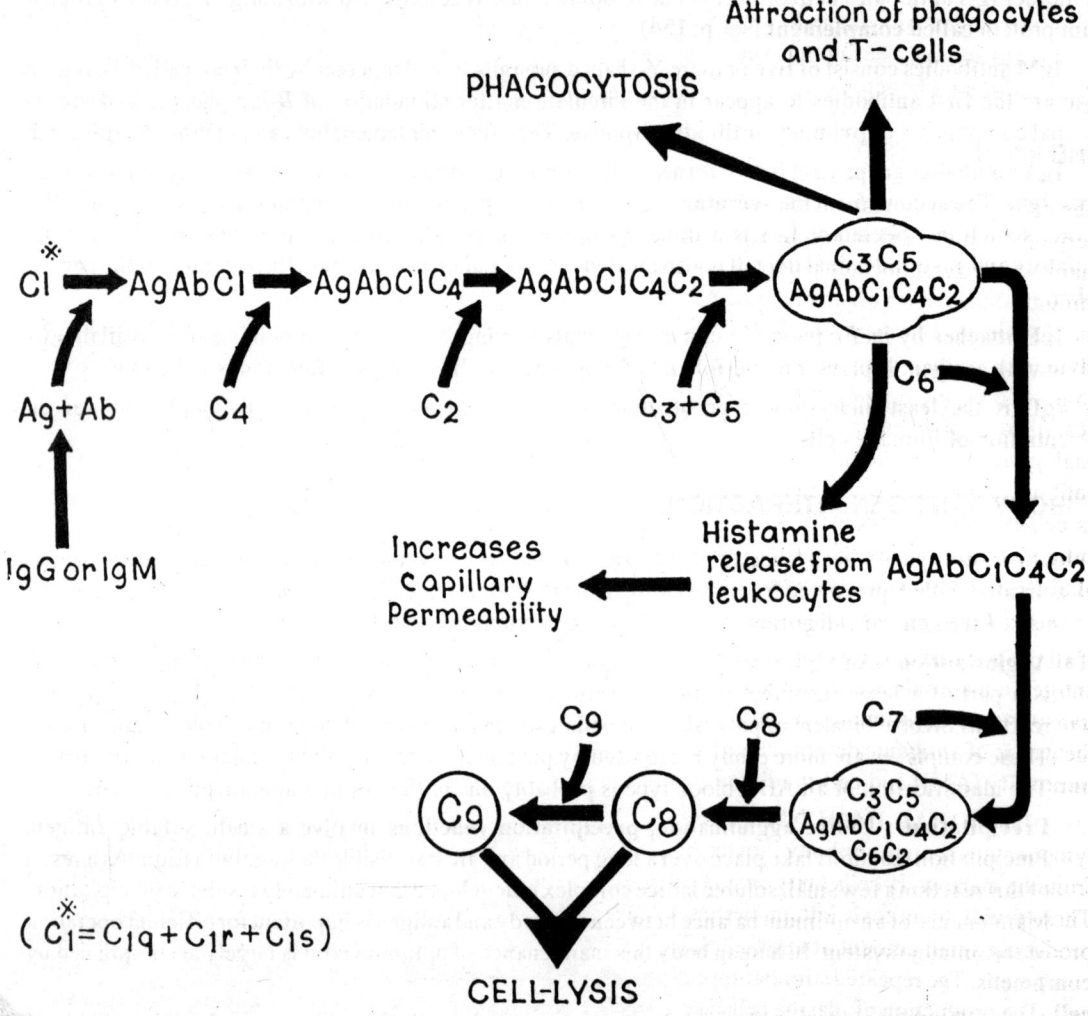

Fig. 11.2 Involvement of complement system for the lysis of bacterial cell.

(4) **Opsonisation** : It is the fourth main antigen-antibody reaction associated with humoral antibodies. Certain antibodies and two complement factors $(C_3 C_5)$ are the main **opsonins** that stimulate phagocytosis. The presence of C_3 and C_5 molecules on the cell surface makes the microbes more recognisable to phagocytes leading to the enhancement of phagocytosis. These complement factors stimulate F cells to move into the area (positive chemotaxis) which begin the process of cell-mediated immunity and causes the release of histamines from leukocytes. The histamine release increases capillary permeability and smooth muscle contraction. Taken collectively these reactions result in local inflammation.

The activation of complement factors does not always follow the general pathway as depicted in figure 11.2. In humans a bypass system called the **properdin system** is able to activate the $C_3 C_5$ complex initiating the same protective responses. The significance of this system is evident in persons showing the infection of *Neisseria meningiditis* that lack an efficient complex system $(C_5 C_9)$.

(5) **Neutralisation** : The final type of antibody-antigen reaction associated with the humoral immunity is known as **neutralisation**. The reaction may occur with soluble toxin molecules released by the pathogen or with individual viruse. The anitbodies that participate in the reaction is known as **antitoxins**. The antigen-antibody complex blocks the reaction site of the toxin neutralising the effect of the pathogen.

THE IMMUNE SYSTEM

The cells primarily responsible for the operation of immune system are the **lymphocytes**. These cells are derived from the **primordial cells** called the **stem cells** in the bone-marrow. On coming from the bone marrow, the lymphocytes enter the lymph system (lymph node, spleen, tonsils, thymus etc). Some pass through a specialised organ of thoracic cavity called **thymus**. In human, this flat, bilobed organ lies just below the thyroid gland, above the heart. Within the thymus, the lymphopoitic cells are differentiated to form thymus dependent lymphocytes or **T-lymphocytes (T-cells)**. The other type of lymphocytes, that are not controlled by the thymus, are known as bursa-equivalent cells or **B-lymphocytes (B-cells)**. This was first described in chicken as a gland, the **bursa of Fabricius** in the lower gastrointestinal tract. In human the analogous organ is thought to be fetal liver or bone marrow. T-cells formed from the thymus do not secrete antibody but aid the B-cell in regulating the production of antibody and play a vital role in a process known as **cell-mediated immunity** (CMI). The T-lymphocytes and B-lymphocytes are able to intercept all the antigens except those entering the cardiovascular system directly. Usually the B-lymphocytes have a life span of about five to seven days whereas T-lymphocytes may live for months or years.

With the entry of antigens into the lymphatic system the immune response begins. The antigens are first of all phagocytized by macrophages, monocyte or polymorphonuclear cells (PMN) (see p. 133). Majority of antigenic materials are destroyed by this defense mechanism. The remaining antigenic determinants are transported to the immune system to be acted upon by the T-lymphocytes or B-lymphocytes. Depending upon the nature of antigenic determinants two forms of immunity are displayed - **humoral** or **antibody mediated immunity** (AMI) and **cell-mediated immunity** (CMI).

HUMORAL IMMUNITY : B-cells play a vital role in the process of AMI. These differentiated lymphocytes secrete antibody into the body fluid (humor) in response to antigenic stimulation. After release from the marrow, B-cells that come in contact with antigen are stimulated to enlarge and become blast cells. These divide by mitosis into numerous smaller lymphocytes capable of responding to the same antigen by producing small amount of antibodies. Cells that are able to respond in this manner are said to be **immuno-competent**. The repeated stimulation and division results in the formation of a clone of cells known as **plasma cell**. The production of plasma cells is under the control of genes, called the **immune response gene** (Ir). In humans the **Ir** genes occur on various chromosomes. Antibodies are synthesised as per the requirement and direction of **Ir** genes that are released into the circulation at a rate several thousand molecules per second.

The plasma cells are the primary antibody producing cells of the body and are responsible for humoral immunity. They continue to produce antibodies for two to three days or till the antigenic determinants are exhausted. At this point the plasma cells die off and are replaced by a second *B*-lymphocytes called **memory-B-cells**. These memory cells remain in lymphoid tissue for many years and become functional when the same antigen reappears producing thereby the specific plasma cell to destroy the same. The production of antibodies and their specific reactions with antigens provide the body with an effective, selective defense mechanism against invading foreign materials and microbes.

CELLULAR IMMUNITY : The second general type of immunity is mediated by T-cells. These cells are programmed for the participation in cell mediated immunity or tissue immunity. T-cells produce no antibody but provide protection in the following manner.

The antigens of many fungi, protozoa, some specific viruses and bacteria (specially intracellular parasite) stimulate the T-lymphocytes and sensitize them to enter the circulation and migrate to the site of invasion. At the antigen site the lymphocytes revert to immature cells called **lymphoblast** which later on produces a series of low molecular weight protiens, the **lymphokines**. These are the expanded clone of immunocompetent lymphocytes that can be functionally defined in three categories : **effector cells, regulator cell** and **memory .cell**. A generalised feature of cell-mediated immunity is presented in table 11.3.

Table 11.3

CELL-MEDIATED IMMUNITY

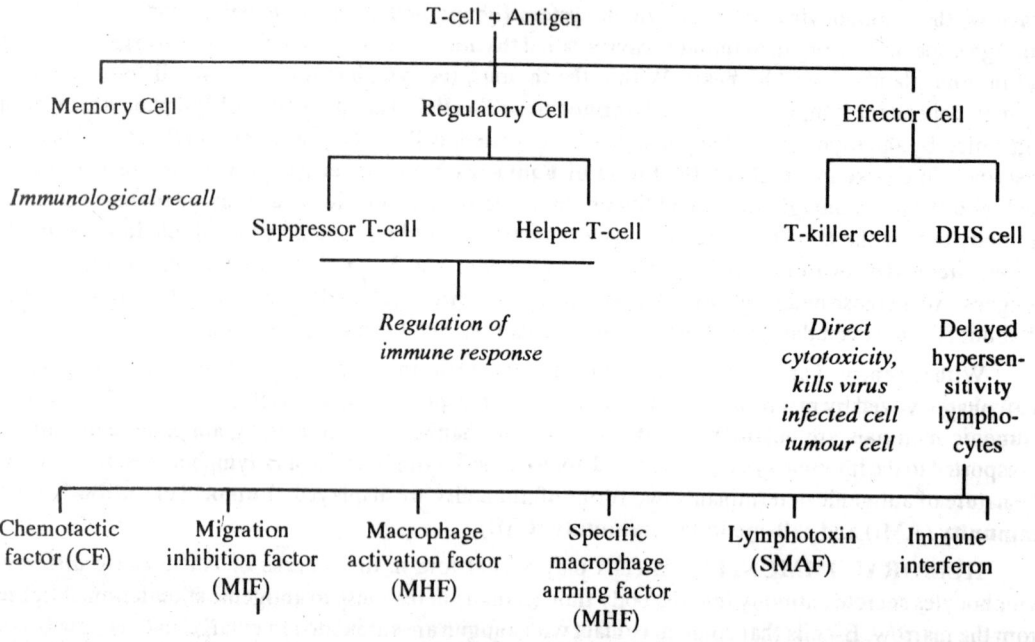

EFFECTOR CELL : Effector cells are the lymphocytes that provide the actual protection. For example, **T-killer lymphocyte** physically attaches to their target cells and destroys them by membrane desruption or lysis. These 'killer' cells are much better adapted to eliminate some tumor cells and viral infected host cells. Other effector cells are **delayed hypersensitivity lymphocytes** (DHS) cells which, though unable to damage directly the cellular target, hasten the process of destruction of foreign cells by their active participation. The lymphokines of DHS that increase the phagocytic efficiency are :

Migration inhibition factor (MIF) - reducing the mobility of macrophages so that they accumulate around the activated T-cells.

Chemotactic factor (CF) : attracting macrophages to area of antigen concentration.

Macrophage activating factor (MAF) : enhancing phagocytosis of foreign particles, creating a population of 'angry macrophages' with larger, more lysosomal granules.

Specific macrophage arming factor (SMAF): enhancing the ability of macrophage to kill the specific antigenic target cells.

Lymphotoxin : destroying non specifically potential pathogenic cells; this has some adverse responses on host tissue also showing allergic effects (discussed later in this chapter).

Immune interferon : aiding the arrest of viral infection by increasing the cytotoxicity of T-cells and triggering the production of antiviral protein in neighbouring cells (refer to p. 158).

The overall effect of these lymphokines is to increase the efficiency of phagocytosis of antigen and bring about a specific response to the disease. CMI defense is usually non-specific and many unrelated pathogens may be simultaneously eliminated by this stimulated system. Recently in 1979, the term **interleukin** was coined for substances produced by leukocytes that influence other (inter) leukocytes. For example, interleukin-1 is a T-lymphocyte protein (lymphokines) that stimulates the maturation of T-lymphocytes; interleukin-2 activates T-lymphocytes to rapidly grow and divide. Interleukin-2 has practical significance in treatment of tumor.

REGULATOR CELL : Regulator cells are T-lymphocytes that govern the intensity of immune response and regulate the co-operation between CMI and humoral immunity. **Helper T cells**, for example, co-operate with B-cells to initiate an antibody response. Another **suppressor T-cells** effect B cells in opposite fashion. They apparently interfere with the B-lymphocytes and prevent and exaggregate immune response that might damage the host. In the AIDS victim an abnormally low number of helper T-cells exist in the immune system together with an unusually high number of suppressor T-cells resulting in the suppression of immune response, the characteristic of the disease.

Table 11.4

A COMPARISON BETWEEN ANTIBODY-MEDIATED AND CELL-MEDIATED IMMUNITY

	Antibody-mediated	Cell-mediated
Sensitizing antigen	Protein, polysaccharide, lipid hapten,	Protein or protein whole cell.
Reaction time	Immediate	Delayed, 18-24 hours.
Initiating event	Union of antibody with antigen	Reaction of sensitized lymphocytes
Transfer	Circulating serum	Cells antibody
Effector mediators	Complement factor	Lymphokines
Benefits	Antimicrobial immunity	1. Antimicrobial imm-to microbes & bacterial unity to virus, toxins bacteria, fungi, protozoa 2. Tumor immunity 3. Transplantation immunity.

Source : J.A. Bellanti, Immunology : Basic Process, 2nd ed. W.B. Saunders Co. 1979.

MEMORY CELL : A third pool of antigen-stimulated T-lymphocytes are **sensitized memory T-cell**. These cells, remain in quiescent stage in the host after the elimination of the invading antigen. They retain the potential for rapid activation whenever the host is exposed to the corresponding antigen. The CMI recall response is similar to the humoral **anamnestic** response that decreases the duration of stimulation and neutralisation phase considerably. This is one of the major factors of long term immunity to disease. A comparative account of humoral immunity and cellular immunity is presented in table 11.4.

NATURAL KILLER CELLS AND INTERFERON

T-lymphocytes, *B*-lymphocytes and macrophages are the major components of immune system. Besides these, some **natural killer** (NK) cells are also present in human body that are able to lyse a variety of virus-infected cells and tumors. A person exhibiting mumps and herpes virus infection accumulates great number of natural killer cells. It is believed that the formation of NK cells is largely regulated by the release of **interferon** from virus infected cell. Moreover, the activity of NK cell is substantially enhanced by the lymphokines, immune interferon.

Interferons are a group of related proteins having molecular weight 20,000-30,000. They are mild antigens. These are non-dialasable, non-sedimentable and non-precipitable by perchloric acid; can withstand temperature upto 60°C and show stability at wider pH range (2.0-7.0). With the first discovery of interferon by Issac and Linderman in 1957 in the allantoic fluid of eggs, the chemical has received general acceptance as antiviral agent and has been suspected as effective control measure for many viral diseases, especially in tumor formation. Interferon are non-specific. Human and monkey interferons elicit cross-reactions. Recently cross-reaction of interferon between more distinct species e.g., human and rodents has also been recognised (Stewart, 1974). It has been found in all human cells in response to certain stimuli. Contact with viruses, rickettsias, protozoa, bacterial exotoxins, certain antibiotics (e.g., cyclohexamide and kanamycin) stimulates the release of small amount of these inhibitory molecules. The most effective interferon producing cells are lymphocytes, fibroblast and T-cells. For experimentation, interferon is extracted from the prepuce tissue of male infants that has been found to be effective against herpes simplex, hepatitis B and common cold infection.

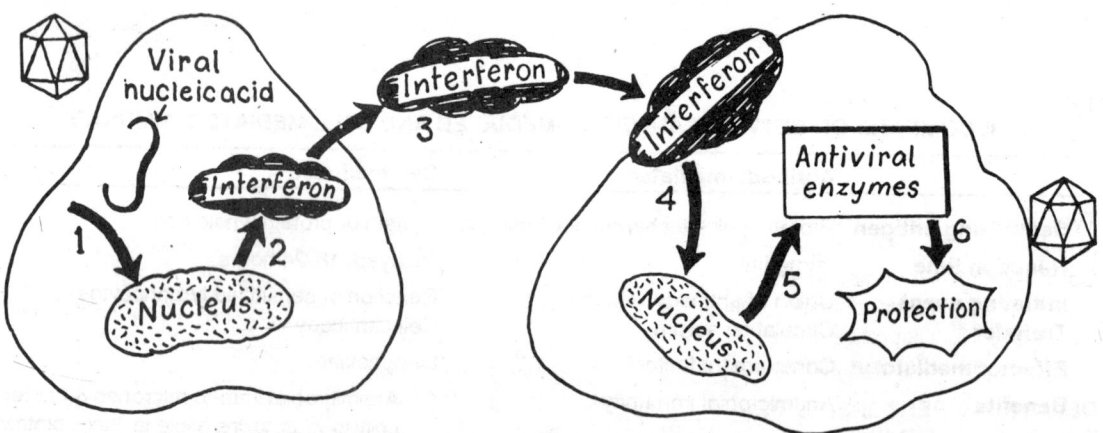

Fig. 11.3 Antiviral mechanism of interferon.

1. Viral nucleic acid turns of host gene for interferon production 2. infected host cell produces interferon 3. interferon diffuses out of infected cell and attaches to the membrane of uninfected cell 4. attachment triggers signal in the nucleus of uninfected cell 5. Signal turns on the production of antiviral enzymes 6. antiviral enzymes prevent the synthesis of virus-specific protein protecting the infection of uninfected cell.

Interferon does not interfere directly on the viral metabolism rather it diffuses from the viral infected cell to other non-infected neighbouring cells and stimulates these non-infected cells to produce enzymes with antiviral property. The antiviral mechanism of interferon is presented in figure 11.3. Recent studies by Swiss and Japanese workers (1980) in deciphering the genetic code for interferon gave a major break-through in the treatment of leukaemia by introducing γ, ß and α interferon in the market under the trade name **intron**. The specific mechanisms attributed to these inhibitors are spelled out as :

  (i) inhibition of viral mRNA translation

 (ii) blockage of viral nucleic acid transcription and

(iii) alteration of host cell membrane.

      Propanediamine is a recently introduced interferon against Rhinovirus (cold) infection. Interferon capable of inhibiting influenza type A virus has also been found to be effective against equine encephalitis and vaccinia. It has opened a new avenue to test the efficacies of many interferons against virus mediated lymphoma and other malignancies in human beings.

HYPERSENSITIVITY

The immune system is an extremely complex defense mechanism. Under normal conditions, the sensitized immune cell to a particular antigen develops the efficiency and accuracy of eliminating the harmful foreign

Table - 11.5
GELL - COOMBS CLASSIFICATION FOR HYPERSENSITIVITY

Class	Example	Antigen	Antibody	Cell involved	Response	Transfer of sensitivity by
IMMEDIATE HYPERSENSITIVITY						
Type I (Anaphy-lactic)	1. Bronchial Asthma 2. Bee sting 3. Drug reaction 4. Skin transplant 5. Food allergy	Usually from outside the body (pollen, fungi, dust)	IgE	Mast cells, Basophils	5-30 minu-tes; wheal formation kinin release Edema fluid	Serum antibody
Type II (cytotoxic)	6. Glomerulonephiritis, 7. Rh disease in newborn 8. Drug reaction	Cell surface	IgG IgM	RBC, WBC Platelets, Complement	Hours to days; cell lysis	Serum antibody
Type III (Immune Complex)	9. Tertiary syphilis, 10. Lepromatous leprosy 11. Rheumatoid arthritis 12. Lymphoblastic Leukaemia 13. Ulcerative colitis 14. Organ rejection	Outside the body	IgG IgM	Host tissue cells, PMN, platelets, Complement	4-8 hours; Acute inflammation, Edema ,Erythema, Cell lysis	Lymphoid cell
DELAYED HYPERSENSITIVITY						
Type IV (Cellular)	15. T.B Skin test 16. Fur, cosmetics, poison 17. Food allergies 18. Transplanted tissue 19. Chronic microbial infection.	Cell surface or outside the body	None	T-cells, Lymphokines, Macrophages Lymphokine necrosis	24-48 hours; Induration, Erythema, Cytotoxic and	Serum antibody

agents. However, sometimes the immune sensitization to antigen is over-reacted by the host causing more harm than good to the body. This condition is known as **hypersensitivity** or **allergy**. Dust particles, pollens, fungal spores, eggs, drugs etc are some of the common **allergens** that stimulate many allergic reactions like asthma, eczema, itching, skin rashes etc and are clinically referred to as immunological disorders.

As with immunity, there are two basic types of hypersensitivity : (i) **immediate type allergy** - dependent on B cell immune system showing allergic responses within minutes after the exposure to an antigen and (ii) **delayed type hypersensitivity** - associated with sensitized T lymphocytes showing reactions 24 to 48 hours after the exposure. In both the cases, the allergic responses are sometimes so virulent that, if left uncared, may cause permanent impairment of tissues leading to death. As knowledge in this area has increased considerably, there have been many attempts to classify hypersensitivity diseases. In the 1970s Gell and Coombs developed a very useful classification system for reactions responsible for hypersensitivities correlating to the immunological disorders (Table 11.5).

TYPE I - (ANAPHYLACTIC HYPERSENSITIVITY)

This type of hypersensitivity is also known as **immediate, reaginic** or **atopic** hypersensitivity. The symptoms of this reaction are primarily associated with the release of immunoglobulin **IgE**, also known as **reagin**. IgE is a bivalent antibody and has the ability to bind to special receptor sites on the surface of **mast cells and basophils**. One of the most frequent type I allergic response develops after being stung by bees, hornet, yellow jackets, fire ants etc.

Mast cells are connective tissue cells found near the blood vessels and in the respiratory and gastro-intestinal tracts. Basophils are a type of circulating leukocytes, rich in granules. B-cells capable of producing

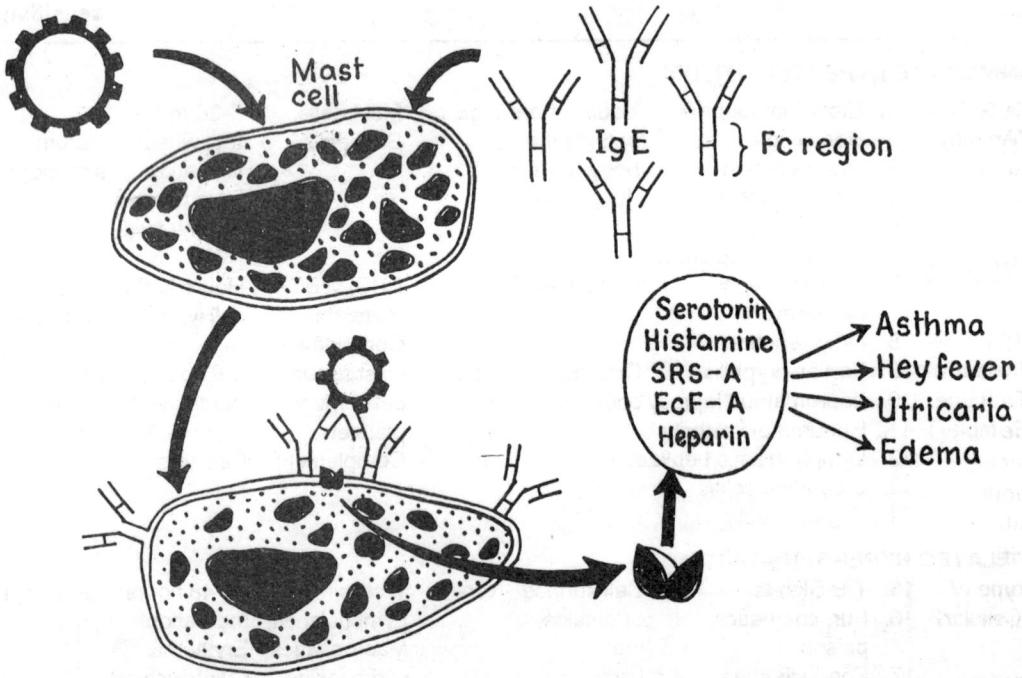

Fig. 11.4 Mechanism of immediate type hypersensitivity.

IgE become sensitized to the allergen when it first enters the body through the skin respiratory tract or gastro-intestinal tract. Upon second exposure, an anamnestic response occurs, releasing additional **IgE** which becomes attached to more mast cells and basophils. Allergen that reacts with this bound antibody stimulates mast cells and basophils to release chemicals that participate actively in membrane distortion and allergic responses (Fig.11.4). The mediator chemicals known as **vasoactive amines**, include varieties of substances like histamine, slow reacting substances of anaphylaxis (SRS-A), eosinophil chemotactic factor of anaphylaxis (ECFA), heparin, serotonin, kinins and postglandins. The release of these amines is responsible for the symptoms of both **generalised** and **localised** anaphylaxis. Generalised anaphylaxis is a shock reaction that occurs after the second exposure to an allergen. The symptoms may vary from a mild reaction to massive response that may be fatal within minutes. The release of vasoamines causes **utricaria** (skin reaction) which ultimately progresses to excessive skin reddening and itching, capillary breakage, burning sensation in the rectum, mouth, quick drop in blood pressure, constriction of bronchial smooth muscle and death. The localised (atopic) anaphylaxis is characterised by many of these symptoms; however, they are chiefly restricted to a particular organ (skin, nasal mucosa, intestinal linings etc). Bronchial asthma involving contraction of bronchiole smooth muscle, hay fever caused by inhalation of grass or weed pollen and food allergies accompanied by cramps, nausea and diarrhoea are the common examples of atopic anaphylaxis. It is believed that an atopic person lacks sufficient IgA lymphocytes to block the allergen or has a defective suppressor T-lymphocyte to neutralise the antigen.

TYPE II - (CYTOTOXIC HYPERSENSITIVITY)

This type of hypersensitivity results in the destruction of host tissue by lysis. The allergens usually come from outside the body (e.g., penicillin, blood transfusion) or may be due to biochemical changes on the host cell surface. The most likely antibodies able to combine with these antigens belong to IgM and IgG classes and contain complement binding sites on the Fc portion of the molecules. The binding of C_{1q} complement to the antibody begins the complement dependent sequence and ends with lysis and inflammation of the host tissue. Cells undergoing complement dependent lysis include red blood cells, brain cells, T-cells etc. Type II hypersensitivity also occurs in conjunctions with drugs (insulin, tetracyclin, antihistamine, penicillin etc) but unlike type I allergies in this case the drug molecules get attached to the surface of the red blood cells and form strong antigens that stimulate the release of antibody and lysis by complement fixation. One of the most common examples of cytotoxic hypersensitivity is evident in **hemolytic disease of the newborn** or **Rh disease**.

Some persons contain Rh antigen (Rh = Rhesus; the antigen first discovered in rhesus monkey) in their erythrocytes. They are said to have Rh-positive blood group and those lacking it are Rh-negative individuals. Production of Rh-antigen is an inherent trait which normally passes from the father's side to the offsprings. If a Rh-negative woman marries an Rh-positive man, there are chances (as per Mendellian law) of receiving the Rh-positive fetus by the woman. During the birth process, the woman's circulatory system is exposed to her child's blood due to the broken placenta membrane. If the child is Rh-positive, its antigen will enter the mother's body and will stimulte her immune system to produce anti Rh(+) antibodies. In case of second or subsequent pregnancy with mother's Rh-positive fetus, these antibodies will also cross the placenta along with other antibodies and enter the fetus circulation. There they will react with the Rh antigen and cause complement mediated lysis of the cells. This will result in **erythroblastosis**. The **Rh-disease**, clinically known as **erythroblastosis fetalis** leads to still birth of the child.

TYPE III - (IMMUNE COMPLEX HYPERSENSITIVITY)

This type of hypersensitivity is largely mediated by IgG and IgM. The symptoms include an acute inflammatory response and tissue destruction in the basement membrane and blood vessels. The reaction is governed by the soluble antibody-antigen complexes that accumulate in the blood vessels or on target tissues. In normal conditions the antigens are effectively phagocytized by the antibodies but quite often the antigens continue to accumulate on the reaction site in excessive amounts resulting in the formation of soluble immune complexes that escape phagocytosis. The concentration of immune complexes stimulates the complement system, the release of kinins, PMN migration, increase in permeability of capillary and the release of histamines. These ultimately cause tissue damage.

One of the most typical and well-studied Type III reactions was first studied in 1903 by Maurice Arthus, what is now known as **Arthus reaction**. After a series of horse serum injection in laboratory animal, the skin of animal becomes sensitized to antigenic horse protein. Further injection of serum results in an Arthur reaction at the injection site. The excess antigen binds with antibody and immune complex accumulate on cell which form the boundaries between the dermis and epidermis. The injection side swells with edema fluid and homolysed red blood cells as inflammation progresses. In severe cases the limited oxygen supply may lead to the development of gangrenous tissue.

Serum sickness and systemic lupus erythrematosus are other human type III hypersensitivity. Serum sickness is a response that occurs in tissues after repeated injection of DPT horse antiserum used to provide acquired passive immunity. Symptoms include fever, malaise, urticaria and enlargement of the lymph nodes and spleen. Serum sickness is also evident in some individuals in response to antibiotic therapy or kidney transplantation. SLE results from the formation of immune complexes of antinuclear antibody and nucleic acid antigen. Since SLE cases have been reported to occur in the same family over a number of generations, many suspect that the disease is influenced by genetic factors.

TYPE IV - (CELLULAR HYPERSENSITIVITY)

Delayed hypersensitivity begins in much the same way as cell mediated immunity. The allergen combines with special surface receptors on sensitized T-cells triggering them to enter a period of blast transformation after which lymphokines and interferon are released into the surrounding tissue. A delay of about 12 hours occurs before the effects of these processes are visible. **Induration** (a hard spot) and **erythema** (reddening) are the typical symptoms of delayed hypersensitivity. There are two major forms of this hypersensitivity : infection allergy or TB skin test reaction and contact dermatitis.

TB skin test reaction is perhaps the most familiar delayed hypersensitivity response to detect the infection of *Mycobacterium tuberculosis*. The test is performed with either OT (old tuberculin) or PPD (purified protein derivatives). The antigen (**tine**) is pressed to the skin intradermally with a hypodermic syringe. Sensitivity to *M. tuberculosis* usually takes place within 4-7 days after an individual has first come in contact with the antigen. Induration and erythema develop gradually and reach a peak between 24 to 48 hours and then slowly disappear. A positive TB skin test shows an area of induration of at least 10 millimetre in diameter with surrounding erythema indicating the patient has been exposed and sensitized to the antigen.

Contact dermatitis is a type IV hypersensitivity that occurs in response to something a person touches. Although many substances may harm the skin, not all dermatitis stimulate immune response. Many household cleansers, detergents, acids, alkali may cause dermatitis; they do not stimulate antibody formation. These are known as **primary irritants**. On the other hand there are a number of materials (plants, metals, natural materials, synthetic materials) that are responsible for true allergic contact dermatitis. Coming in contact with these allergens results in rashes, itching, blisters (vesiculation), urticaria. In common parlance these symptoms are referred to as **eczema** which varies in severeity from individual to individual. Metals, cosmetics, hair dyes etc are the common dermatitis allergens that induce varying symptoms in the human beings. Contact dermatitis rarely results in severe permanent tissue damage. However, delayed hypersensitivity occuring after tissue transplant may result in direct cell-mediated cytotoxicity.

SEROLOGY

Laboratory techniques that specifically depend on antibody-antigen reactions for the detection of diseases or infection are known as **serological test**. The test primarily aim at demonstrating the presence of an elevated blood concentration of antibody that specifically reacts with the microbial antigens or its extracellular products. Presence of elevated antibody titers indicate that a person has been exposed to the pathogen and has responded immunologically. The information also offers indirect evidence of the etiology of diseases, often before the pathogen can be isolated and identified. In cases, where cultivation of fastiduous or delicate pathogen is difficult or impossible (e.g., syphilis) serological findings become increasingly important. The test is more relevant in case of serious diseases that progress slowly. For diseases of short duration, the test is, however, of limited therapeutic value because immunological system usually requires several days to respond to the antigen. Serological tests are also helpful in identifying the analysis of unknown specification.

Most serological tests require a source of antigen. However, under certain circumstances, when microbial antigens are not readily available, analogous substances of similar antigenic specificities are substituted for the natural antigens. The antibody may cross-react with the heterophile antigens producing the identical immunological response, as is evident in many viral infections. The methods of detecting the specific antibody-antigen reaction, however, differ considerably for different diseases. Agglutination, precipitation, neutralisation are the observable activities that can be used to detect these positive reactions (refer p 153). Besides, some special sophisticated tests are used in serology for the detection of specific antigen. Some of these commonly employed serological tests are mentioned below :

1. Complement fixation test

The test is used in the identification of many diseases such as mycoplasmal pneumonia, Q fever, polio, rubella, histoplasmosis, streptococcal infection etc. The test utilises a patients's serum, the test antigen,

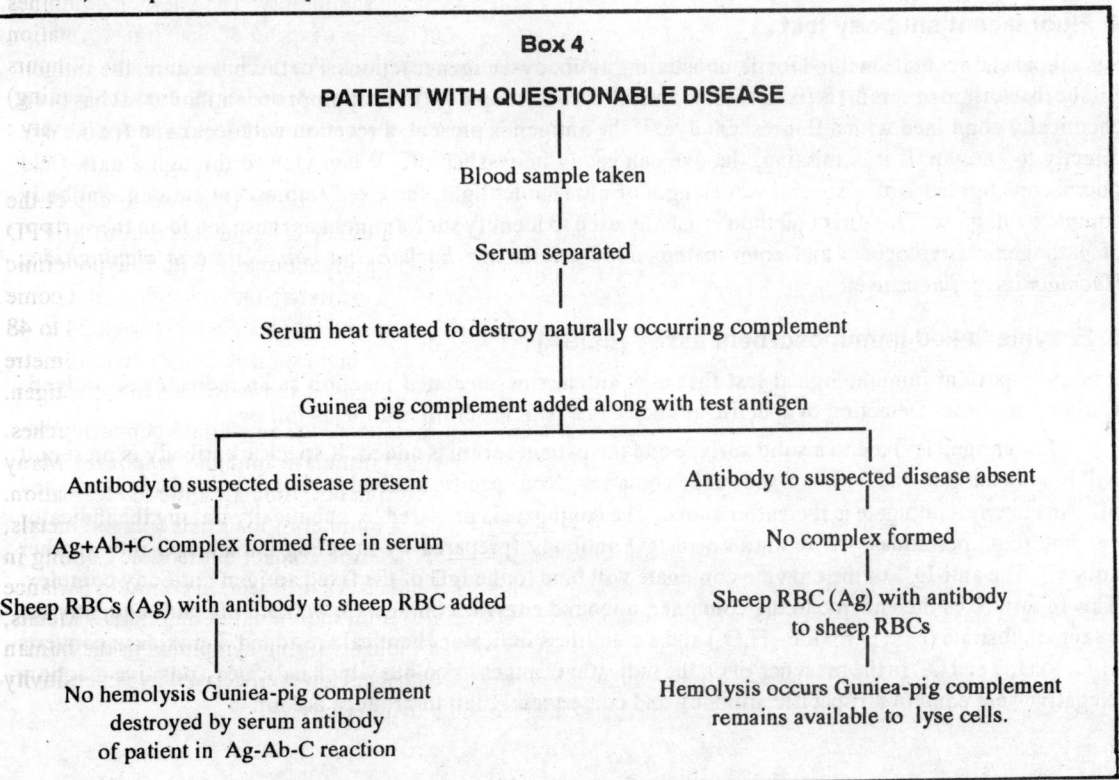

Box 4

PATIENT WITH QUESTIONABLE DISEASE

Blood sample taken

Serum separated

Serum heat treated to destroy naturally occurring complement

Guinea pig complement added along with test antigen

Antibody to suspected disease present	Antibody to suspected disease absent
Ag+Ab+C complex formed free in serum	No complex formed
Sheep RBCs (Ag) with antibody to sheep RBC added	Sheep RBC (Ag) with antibody to sheep RBCs
No hemolysis Guniea-pig complement destroyed by serum antibody of patient in Ag-Ab-C reaction	Hemolysis occurs Guniea-pig complement remains available to lyse cells.

complement from guinea pigs and antiboy to sheep red blood cells to determine whether or not the sheep red cells is lysed by guinea pig complement. If no hemolysis occurs, the antibody to the suspected disease is present and the patient has the illness. If hemolysis occurs, the antibody to the suspected disease is absent and the patient does not have the disease (see Box 4)

2. Flocculation test

Flocculation reactions occur when the antibody-antigen combination separates from solution in clearly visible particles rather than in the more solid mass typical of precipitation or agglutination. The tests are usually run to identify the presence of serum antibody associated with diseases such as syphilis, hepatitis, infection mononucleosis etc. One of the most common flocculation tests is known as the venereal disease research laboratory test **VDRL**. A positive VDRL test shows a high or increasing titer of a serum antibody and is characteristic of most individuals with syphilis infection. Syphilitic patients release a compound called **cardiolysin** which stimulates the immune system to produce reaginic antibody.

3. Antistreptolysin O Test

Popularly known as ASTO test, it is used to identify the presence of a serum antibody produced against the strongly antigenic streptococcal product streptomycin O. This is the most commonly performed serological test for the identification of rheumatic fever. The release of streptolysin O during some streptococcal infection (e.g., septic sore throat, scarlet fever) stimulates the production of antistreptolysin O and a second antibody (autoantibody) that is able to react with cardiac tissue. It is the presence of the autoantibody that results in rheumatic fever symptoms such as heart muscle and valve damage, fever, inflammation, scarring and anemia.

A high ASTO may indicate the presence of damaging autoantibody and helps to differentiate between streptococcal infection and other diseases with similar symptoms.

4. Fluorescent antibody test

It is a rapid and accurate method for demonstrating antibody-antigen reactions. For this procedure, the antigen (tissue, bacterium or serum) is fixed to slide and allowed to couple with the appropriate antibody that has been chemically combined with a fluorescent dye. If the antigen is present, a reaction will occur and fix the dye directly to antigen. If it is missing, the dye can easily be washed off. When viewed through a dark-field microscope lighted with a special wavelength of ultra-violet light, the fixed fluorescent antigen -antibody complex will glow. This direct method is usually used to identify such antigens as those found on the surface of pathogenic streptococci and enteropathogenic bacteria like *Escherichia coli, Neisseria meningitidis, Haemophilus influenzae* etc.

5. Enzyme linked immunosorbent assay (ELISA)

It is an important immunological test that uses an enzyme-mediated reaction as an indicator of antigen-antibody reaction. Detection of specific antibody in a patient serum is done as follows :

The antigen is fixed to a solid surface and the patient serum is added. If specific antibody is present it will bind to antigen and will be fixed to the container. Non specific antibody will fail to bind and is washed off. An enzyme conjugate is thereafter added. The conjugate is prepared by chemically linking the indicator enzyme (e.g., peroxidase) with antihuman IgG antibody (prepared by injecting human IgG into another animal). The anti-IgG of the enzyme conjugate will bind to the IgG of the fixed antigen-antibody complex. This in turn fixes the enzyme to the container, unbound enzyme conjugate is washed off. Subsequently the enzyme substrate (e.g., peroxide - H_2O_2) and a colourless indicator chemical are added. Peroxidase converts H_2O_2 to H_2O and O_2. In the presence of O_2 the indicator changes its colour which indicates a positive reaction. Negative sera contain no specific antibody and consequently fail to produce colour.

6. Radio immuno assay (RIA)

It is a highly specific and sensitive method for detecting minute amounts of antigens of low molecular weight, hapten. This method is useful to identify the presence and amount of antibodies that differ in only a single amino acid in tissues and fluids. The presence of specific antibody in tissue is identified by using radioactive iodine (^{125}I) coupled to an antigen as a marker. If the molecule in question is present, radioactivity may be detected by placing a photographic film over the surface of the test slide and taking a picture of the molecules. The presence of white spots on the film indicates the location of the radio-labeled molecules, while the intensity of the spots may be used to determine the amount of labeled antibody in the sample.

7. Radio allergosorbent test (RAST)

Like RIA this test is also used to detect IgE or other antibodies as well as varieties of small antigens. To detect the IgE, specific antigens are attached to a matrix particle. Serum suspected to contain IgE is then added. Antibody, if present, combines on the surface of the particles. Now an antiglobulin carrying a radioactive label is added to the system. The entire complex will therefore be radioactive,, if the antiglobulin antibody combines with the IgE. If IgE is not present, the particles will not show radioactivity.

SUGGESTED READINGS

Bellanti, J.A. (1979). Immunology : Basic Process 2nd ed. W.B. Saunders Company, Philadelphia.

Burke, D.C. (1977). The Status of Interferon. Scientific American.

Cooper, M.D. and A.R. Lawton (1974). The development of Immune System. Scientific American.

Gell, P.G.H; R.R.A. Commbs and P.J. Lachman (eds). (1975). Clinical aspects of Immunology 3rd ed. F.A. Davis, Philadelphia.

Holland, J.J. (1979). Interferon I : On the threshold of clinical application. Science, 204.

Interferon II : Learning how it works. Science, 204.

Patterson, R. (ed) (1980). Allergic diseases : Diagnosis and Management. J.B. Lippincot, Philadelphia.

Rose, N.R. and H. Friedman (1980). Manual of Clinical Immunology. American Society of Microbiology, Washington, D.C.

Ross, F.C. (1983). Introductory Microbiology. Charles E. Merrill Publishing Company. A Bell & Howell Company, Columbus, Toronto.

GLOSSARY

Abscess - accumulation of pus in a localised area.

Acid-fast stain - differential stain for detecting bacteria that retain carbolfuchsin when treated with acid alcohol.

Acquired immunity - protection from disease due to antibodies or immune cells acquired during a person's life time.

Acquired immun deficiency disease - loss of immunological defence caused by transmissible agents (eg., AIDS).

Active immunity - production of antibodies or lymphocytes by the host in response to the presence of an antigen.

Aflatoxin - secondary metabolites produced by *Aspergillus flavus* or *A. parasiticus*, which are toxic to mammals.

Agammaglobulinema - absence of antibody production due to a defect in B-cell system.

Agglutination - clumping of cells or particles in the presence of specific antibody.

Agranulocyte - white blood cells that lack observable intracellular granules (e.g., lymphocytes, monocytes).

Allergen - antigen that elicit an allergic response.

Allergy - antigen-triggered immune-mediated response harmful to host.

Amensalism - a form of symbiosis in which one species is harmed but the other is neither benefitted nor harmed.

Ames test - laboratory procedure using auxotrophic bacteria to screen mutagenic compounds.

Amination - addition of an amino group (NH_2) to a compound.

Ammonification - release of ammonia (NH_3) from nitrogen containing organic compounds.

Amphitrichous - form of polar flagellation having tufts at each end of the cell.

Anamnestic response - immunological memory that accounts for rapid immune reactions following antigen exposure.

Anaphylaxis - Ig-E mediated allergic response, may be fatal.

Anemia - deficiency in erythrocytes or hemoglobin in blood.

Antagonism - reduction in the effectiveness of one partner by other.

Antibiotics - chemical substances produced by micro-organisms that in low concentrations inhibit or kill other micro-organisms.

Antibody - protein produced in the body in response to an antigen.

Antibody titer - concentration of an antibody in the blood.

Anticodon - three-nucleotide region on tRNA that binds to the codon of mRNA.

Antigen - substance that stimulates the formation of antibody inside the body.

Antigenic determinant - small specific site on the antigen that determines the immune response.

Antigenic drift - minor structural changes in the virus surface due to mutation in the viral genome.

Antigenic shift - sudden major structural change in the virus surface usually due to recombination.

Antiserum - blood serum that contains antibodies of known specificity.

Antitoxin - antibody that neutralises a toxin.

Archaebacterium - group of microbes that lack true nucleus and membrane-bound organelles having some chemical and metabolic properties not found in prokaryotes.

Artifact - artificial characteristic that develops in a specimen as a result of manipulation.

Artificial active immunity - product of antibodies or lymphocytes by the host in response to vaccination.

Artificial passive immunity - immunity acquired by injecting presynthesised antibody into a non immune host.

Asymptomatic - exhibiting no symptom.

Attenuation - less virulent.

Autoimmune disease - disease caused by an immune response against one's body tissues.

Autotroph - organism that uses carbon dioxide for manufacture of food.

Auxotroph - mutant that requires an additional nutritional requirement.

Bacteriological membrane filter - filter that contains pore small enough to prevent the passage of bacteria.

Bacteriophage - virus that attacks bacteria.

Base analog - chemical that resembles one of the base pairs of nucleic acids (purine or pyrimidine), usually causing gene mutation.

Basophil - nonphagocytic granulocyte that stains with basic dyes.

Batch culture - culture practice in which nutrients are neither added nor removed from the container during the growth of culture.

Bergey's Manual - a published classification of bacteria used as a practical guide for identification of bacteria.

B-hemolysis - zone of cleaning around colonies of microbes growing on blood agar.

Bioassay - determining the effect of a substance on a living organism.

Biochemical oxygen demand (BOD) - amount of oxygen consumed by aerobic organisms, used to determine the degradation of organic matter.

Biofertilization - soil enrichment as a result of microbial activity.

Biogenesis - concept that living organisms arise only from living parents.

Biological control - the use of predators, parasites or other biological agents to inhibit or kill unwanted organisms.

Biological indicator - measurement of viability or activities of micro-organisms in particular process.

Biomining - use of micro-organisms to recover metals from ores.

Biotype - subgroup of organisms within a species that possess biochemical properties distinguishing them from other members of the species.

Bivalent - property of an antibody to combine with two identical antigens at the same time.

Blocking antibody - antibody able to bind with an allergen to prevent an allergic response.

B-lymphocyte- cell that has been programmed to participate in the humoral immune response when stimulated by antigen to differentiate into antibody producing plasma cells and memory cell.

Booster - immunization given to enhance the memory response to an antigen.

Cancer - uncontrolled proliferation of abnormal host cells.

Capsid - protein coat surrounding the nucleic acid of a virus.

Capsomere - repeating protein unit of which capsid is formed.

Carcinogen - any agent that causes cancer.

Caseation - dead tissue with a cheese-like appearances typically seen in tuberculosis.

Cell-mediated immunity (CMI) - immune response mediated by T- lymphocytes.

Chancre - ulcer with a hard, rubbery base (e.g., syphilis).

Chemoheterotroph - organism that uses organic chemicals as source of energy and carbon.

Chemoautotroph - organism that uses inorganic chemicals as a source of energy and CO_2 as the sole source of carbon.

Chemotaxis - movement of an organism in response to chemical stimulus.

Chemotherapeutant - chemical for treatment of disease.

Chemotherapy- treatment of disease by chemicals.

Chemotroph - organism that uses chemicals to secure energy.

Chronic disease - slowly progressing disease for long duration.

Clone - group of genetically identical cells that have descended from a common parent cell.

Codon - triplet base sequence of nucleotide in mRNA that helps in the insertion of specific amino acid in protein chain.

Commensalism - association between organisms in which one is benefitted and the other is neither benefitted nor harmed.

Communicable disease - disease that may be transmitted from infected host to uninfected host.

Complement - group of serum proteins that facilitate phagocytosis and lysis.

Complement fixation - reaction of complement with an antigen - antibody complex.

Congenital infection - infection acquired right from the birth.

Contagious disease - spread of infectious disease from person to person.

Continuous culture - a culture system in which micro-organisms are allowed to grow exponentially.

Convalescence - recovery from disease.

Cross-reaction - reaction between an antibody and a diferent antigen than that which stimulated its formation (due to similarities between two antigens).

Cubic symmetry - sphere-like structure.

Cytopathic effect - damage or death of a host cell resulting from virus infection.

Dane particles - infectious form of hepatitis B virus in serum.

Deamination - removal of amino group (NH_2) from a compound.

Degranulation - release of lysosomal contents into the phagocytic vacuole.

Delayed hypersensitivity lymphocyte (DHS) - T cells that enhances immunity by screening lymphokines.

Delayed hypersensitivity - allergic response produced by the cell-mediated immune system.

Denitrification - metabolic conversion of nitrate to nitrite or nitrogen gas.

Differential count - percentge of each kind of leukocyte in a blood sample.

Differential medium - a specific medium used to culture a micro-organism based on its typical characteristic.

Dimorphism - existing in two structurally distinct forms.

DNA homology - degree of similarity between the nucleotide sequence in different DNAs.

DPT vaccine - immunizing mixture against diphtheria, pertussis and tetanus.

Droplet nuclei - expelled particles of moisture that contain micro-organisms and may remain suspended in air.

Early protein - protein synthesised by a virus during the early period of replication.

Eclipse phase - time between virus entry into the host and maturation of the virion.

Edema - accumulation of fluid in tissue.

Effector cell - T-lymphocyte that provides protection from an antigen.

Enzyme-linked immuno-sorbent assay (ELISA) - immunological test that involves enzyme mediated reaction as an indicator of antigen-antibody reaction.

Encephalitis - inflammation of brain.

Endemic - occurring with a constant frequency in the population.

Endocarditis - inflammation of the lining of heart.

Endotoxin - toxin released by the microbe after the death of the organism.

Enriched medium - special medium supplemented with specific nutrients for the growth of particular micro-organism

Enteric - pertaining to intestine.

Enterotoxin - exotoxin affecting the intestinal mucosa.

Eosinophil - non phagocytic granulocyte that stains with acid dyes.

Erythrocyte - Red blood cell.

Etiologic agent - specific cause of disease.

Eutrophication - accumulation of organic matter in an aquatic environment and subsequent increase in microbial populations.

Exotoxin - toxin secreted by microbes in its surroundings.

F⁺ cell - donor cell in bacterial conjugation, contains the fertility factor.

F⁻ cell - recipient cell in bacterial conjugation.

Fibrinolysin - Enzyme that dissolves fibrin clots.

Flagellin - non-contractile protein from which prokaryotic flagella are composed.

Fluorescent antibody technique - method to detect an antigen by treating it with specific antibody conjugated to fluorescent dye.

F-prime cell - bacterial cell with a plasmid that contains some information from the host chromosome and some from the F factor.

Frame-shift mutation - changes in nucleotide base that alter the reading of the frame.

G+C content - percentage of guanine and cytosine nucleotide in DNA, used to determine similarities between organisms.

Gangrene - death of tissue due to loss of blood supply.

Gene amplification - increase in the number of copies of a specific gene in the cell.

Genetic code - triplet nucleotide sequence that specifies a particular amino acid.

Genetic engineering - manipulation of cell's genetic composition and its expression.

Germ theory - theory that infectious diseases are caused by microbes.

Gumma - rubbery necrotic lesion characteristic of teritary syphilis.

Hapten - small molecule that combines with a specific antibody but is non-antigenic unless coupled to a larger carrier molecule.

Helical symmetry - resembling long rod.

Helper T cell - T lymphocyte that co-operates with B lymphocyte to initiate an antibody response.

Hemagglutination - clumping of red blood cell caused by antibodies.

Hemolysis - substance that destroys red blood cells.

Heterocyst - structure in some cyanobacteria known for nitrogen fixation.

Heterotroph - organisms that require carbon in the form of organic molecules.

Hfr cell - donor bacterium containing an F factor integrated into the chromosome which can transfer information to F cell by conjugation.

Humoral immunity - immunity due to the production of antibodies in blood.

Humus - organic matter in soil formed from decayed plants.

Hybridoma - cell arisen from the fusion of a tumor cell with an antibody - producing cell; it produces large quantities of monoclonal antibody.

Hypersensitivity - immune-mediated response to antigen that injures the host (allergy).

Icosahedral - Spherical structure with 20 traingular sides.

ID50 - number of microbes required to cause disease in 50 percent of laboratory animals.

Immediate-type hypersensitivity - allergic response mediated by IgE antibodies, symptoms appear within minutes of the exposure to antigen.

Immunity - a specific host defence mechanism involving the sensitization of lymphocyte to an antigen.

Immunodeficiency - deficiency in either T-cell or B-cell immunity.

Immunopotentiator - substance that enhances the natural immune response.

Immunosuppressant - substance that inhibits the immune response.

Inclusion body - site within the cell containing aggregates of developing viruses.

Indirect transmission - transfer of an infectious agent through an intermediate (vector).

Inflammation - non-specific host response to tissue injury.

Interferon - proteins released by animal cells in response to viral infection that enhance the immune response.

Invasive pathogen - microbes capable to penetrate into deep tissues and disseminate to secondary site in the body.

in vitro - occurring in test tube or culture (in laboratory condition).

in vivo - occurring in a living organism (in natural condition).

Jaundice - yellowing of the skin due to increased bile pigments in blood.

Koch's postulales - criteria used to establish the pathogenicity of a disease.

Lag phase - initial stage in the microbial growth curve (not increasing in number).

Latent infection - infection stage in which a pathogen remains in the host for considerable duration without producing disease symptoms.

Late protein - protein synthesised by a virus during the late period of replication mostly for structural components.

Leukaemia - a state of cancer having uncontrolled proliferation of leukocytes (WBC) and their precursors.

Leukocidin - substance that kills white blood cells.

Leukocytosis - increase in the number of circulating leukocytes.

Leukopenia - decrease in the number of circulating leukocytes.

L-form - bacteria that lack cell wall.

Logarithmic (log) phase - stage in the growth curve showing increase in the number of cells in exponential rate.

Lophotrichous - arrangement of polar flagella in tufts at one end of cell.

Lymphocytes - type of agranulocyte - B-cells and T-cells.

Lyophilization - rapid dehydration of organisms in frozen state.

Lysogenic cells - bacterium that contains a prophage in its chromosome.

Lysogenic conversion - acquisition of new properties by a cell following lysogeny.

Lysosome - intracellular containing hydrolytic enzymes and other oxidizing chemicals.

Lytic cell - virus infection of bacterium that results in lysis.

Macrophages - phagocytic white blood cells that develop from monocyte.

Malignant - invasive tumors that tend to metastasize.

Memory cell - antigen sensitive B or T lymphocyte present in the host that differentiate immunologically active T cells or plasma cells against the particular antigen on subsequent exposure.

Meningitis - inflammation of the brain or spinal chord.

Metastasis - spread of disease from its primary site to another part of body.

Microaerophile - organisms that require lesser amount of molecular oxygen than that in normal air for growth.

Microtubule - hollow protein filaments in eukaryotic flagella and cilia.

Minimum inhibitory concentration (MIC) - the smallest concentration of chemical that inhibits the multiplication of the pathogen.

Minimum medium - medium that supplies only a source of carbon, nitrogen, inorganic salts and energy.

Mixed culture - culture containing more than one organisms.

Molecular biology - a branch of science that deals with the kinds of molecules found in living cells, their behaviour and function.

Monoclonal antibody - immunoglobulin (antibody) specific for a single antigenic determinant produced *in vitro* by hybridomas (lymphocytes fused with tumor cells).

Monocyte - mononuclear phagocyte from which macrophages differentiate.

Monotrichous - single polar flagellum.

Mutualism - symbiotic relationship that benefits both the partners.

Mycorrhiza - fungi in the roots of plants.

Mycosis - disease caused by fungi.

Mycotoxin - secondary metabolites produced by fungi which are toxic to animals.

Mycotoxicoses - diseases caused by ingestion of mycotoxin.

Natural passive immunity - immunity acquired by transfer of maternal antibodies across the placenta to the fetus.

Natural killer cell - protective cell that lyses virus infected cells and tumor cells.

Negative staining - technique that stains the background leaving cells colourless.

Neutrophil - phagocytic granulocyte that does not stain with either basic or acid dyes.

Nitrification - conversion of nitrogen in ammonia to nitrates and nitrites.

Nonsense codon - codon for which no amino acid is specified and that signals the termination of protein chain.

Noscomial infection - infection within the hospital environment.

O antigen - a surface antigen that is component of lipopolysaccharide in Gram-negative bacteria.

Oncogene - gene responsible for transforming normal cells to malignant cells.

Oncogenic virus - virus that can cause cancer.

Operon - genetic unit consisting of an operator site and the adjacent structural genes which are regulated as unit.

Opportunistic pathogen - pathogen that becomes active and produces diseases on onset of favourable condition.

Opsonin - serum protein that promotes phagocytosis of the antigen.

Papilloma - cutaneous tumor e.g., warts.

Passive immunity - transfer of antibodies from an immune host to another individual.

Pasteur effect - preference of facultative anaerobes to use aerobic pathways when molecular oxygen is available.

Peristalsis - rhythmic muscular contractions to expel substances through the digestive tract.

Peritrichous - arrangement of flagella over the entire surface of a bacterium.

Phage typic - identification of bacterial strains using bacteriophage susceptibility as specific indicator.

phagocytosis - cellular engulfment of solids.

Photoautotroph - organisms utilising light as energy source and CO_2 as carbon source.

Photohetrotroph - organisms utilising light as energy source and organic molecules as its carbon source.

Pilus - small tubes extending on the surface of bacteria used for attachment and conjugation.

Pinocytosis - engulfment of liquids by a cell.

Plaque - a clear spot on the surface of cell culture resulting from the lysis of the host cells by virus.

Plasma cell - antibody - producing cell derived from B lymphocyte.

Plasmid - small, circular piece of extrachromosomal DNA in bacteria.

Pleomorphic - exhibiting several different shapes.

Predisposing factors - conditions that make a host susceptible to infection.

Prions - infectious protein complex that lack nucleic acid.

Prokaryote - cell whose genetic material is not surrounded by a nuclear membrane.

Prodromal period - earliest stage of developing infection or disease.

Promoter - site on DNA to which messenger RNA polymerase binds.

Prophage - DNA of temperate bacteriophage that has integrated into host chromosome and established lysogeny

Pus - accumulation of dead white blood cells, bacteria and serum fluid in tissues.

Quarantine - isolation of individual having communicable disease.

Radioimmune assay (RIA) - immunological test that uses a radioactively labeled antigen or antibody to detect its complementary substance.

Recalcitrant - presence of compounds in the environment because of the inability of microbes to degrade them.

Recombinant DNA technology - genetic engineering technique that restructure the cellular DNA producing unique characters.

Refractive index - bending of light through the medium of different density, enhancing the magnificaiton of microscope lense.

Regulatory gene - gene that specifies the production of a repressor protein.

Resistance factor - transmissible plasmid that provides resistance to bacteria against antimicrobial agent.

Resistance transfer factor - portion of R factor that allows the plasmid to be replicated and transferred.

Restriction endonuclease - enzyme that cleaves DNA at specific nucleotide sequence often leaving 'sticky ends'.

Reticulo-endothelial system - a functional system of fixed phagocytes (macrophages) found in association with the endothelium of blood vessel, which are important in inflammatory response and immune response.

Reverse transcriptase - enzyme that polymerises DNA molecule from RNA.

Reye's syndrome - acute, frequently fatal condition of brain and liver.

Sepsis - presence of pathogenic micro-organisms in the blood or other tissue.

Septicemia - presence of actively proliferating micro-organims in blood.

Serology - study of antigen-antibody reactions *in vitro.*

Serotype - strain of microbe that is immunologically distinct from other strains.

Serum - liquid portion of blood that is left after removal of erythrocyte and clotting factor.

Siderophore - compounds that bind with iron.

Sputum - secretion of lower respiratory tract.

Starter culture - inoculum for initiating fermentation.

Sticky ends - single-stranded ends of a linear piece of double-stranded DNA.

Structural gene - gene that specifies the production of an enzyme.

Symbiosis - close association of two dissimilar organisms living together.

Syndrome - complex symptoms of a disease.

Synergism - enhancement of effectiveness of two partners when used in combination.

Target organ - body site most commonly attacked by a particular pathogen.

Temperate bacteriophage - virus that establishes lysogeny with its host bacterium.

Thymus - organ that programmes lymphocytes to become T-cells.

T killer - T-cell that physically attaches to cells and destroys them by lysis.

T-lymphocyte - lymphocytes produced by thymus for cell-mediated immunity.

Toxoid - inactivated form of toxin that is antigenically identical to the active toxin. -

Transcription - process of assembling a molecule of messenger RNA with a nucleotide sequence complementar to corresponding segment of DNA.

Transduction - bacteriophage-mediated gene transfer from one bacterium to another.

Transformation - transfer of genetic information by free DNA released from disrupted bacteria.

Translation - stage of protein synthesis that involves the pairing of mRNA with tRNA and results in alignmen of proper amino acids.

Urethritis - inflammation of the urethra.

Vaccine - preparation of toxoid that can stimulate immunity against a pathogen or toxin.

Vector - living organism that transmits disease from one individual to another.

Venereal disease - infectious disease acquired by sexual intercourse or genital contact.

Vertical transmission - passing out of disease from parent to offsprings.

Viremia - presence of virus in blood.

Virion - infectious virus particles.

Viroid - group of small RNA molecules that cause diseases of plants.

Wandering macrophages - phagocytic cell that travels to site of infection or inflammation.

Weil-Felix reaction - serological test for diagnosing rickettsial disease.

Wild type - organism with non mutant genotype.

Zoonosis - disease transmitted from an animal to human.

ORGANISM INDEX

SUBJECT INDEX